Energy Economics and Policy 1

ENER – The European Network
for Energy Economics Research

Francis McGowan (Ed.)

European Energy Policies in a Changing Environment

With 28 Tables

Physica-Verlag

A Springer-Verlag Company

Francis McGowan
School of European Studies
University of Sussex
Falmer, East Sussex
United Kingdom BN1

ISBN 3-7908-0951-9 Physica-Verlag Heidelberg

Die Deutsche Bibliothek – CIP-Einheitsaufnahme
MacGowan, Francis:
European energy policies in a changing environment : with 28
tables / Francis McGowan. – Heidelberg : Physica-Verl., 1996
  (Energy economics and policy; Vol. 1)
  ISBN 3-7908-0951-9
NE: GT

SPIN 10541032   88/2202-543210 - Printed on acid-free paper

Preface

Across the European Union, energy policy remains a highly controversial issue, despite the relative stability of energy markets and the lack of concern of energy availabilities and prices which marked previous debates on the issue. The debates on nuclear power in Germany, on the coal industry in the UK indicate the continuing resonance of energy policy as a source of dispute at the national level. Moreover, while the urgency which the 1970s energy crises brought to policy discussions may no longer exist, the emergence of new issues, such as the environment and economic liberalisation, have offered new challenges for energy policy.

These issues also arise in a European setting. Early in 1995 the European Commission published a Green Paper on Energy Policy - indeed, as the manuscript was completed the Commission was finalising a White Paper - and the issue will be on the agenda of the 1996 Intergovernmental Conference. The outcome of these discussions is far from clear, but there is no doubt that, in one way or another the European Union will help to shape national energy policies for many years to come.

This book aims to review the development of energy policy at both the national and the European levels. On the basis of a series of country studies, the book aims to show how policies have evolved in the post war period (especially the last twenty years). It outlines how governments' roles have changed in the energy sector and assesses the impact of new policy challenges. The interaction between national and European authorities is also considered, along with a more detailed consideration of the problems which European energy policy encounters.

The book has been written by a team of energy policy experts drawn from the European Network for Energy Research (ENER). This network, established for over a decade and funded by the European Commission, brings together energy policy analysis groups from Belgium, Denmark, France, Germany, Italy, the Netherlands, Portugal, Spain and the United Kingdom. The chapters here have been written by members of the longest established institutes: the Fraunhofer Institute for Innovation and Research, Karlsruhe; the Institute for Energy Economics and Policy, Grenoble; the Institute for Energy Research, Milan; and the Science Policy Research Unit, Sussex. This book is intended as the first in a series of ENER publications.

Although the contributors have debated these issues over many years, there is no common view on how energy policy should develop at the national or the European level. For example, amongst the contributions it is possible to identify those concerned about the absence of a Common European Energy Policy and those who are happy to see such an absence persist. Yet all the contributions strive to inform the debate on energy policy in Europe at both a national and an EU level.

The authors would like to thank their colleagues, both in the ENER network as a whole and in their individual research institutes, for providing a stimulating intellectual environment in which to develop these ideas. The editor would like to thank the authors and colleagues for their support throughout the preparation of this volume. In particular, he would like to thank Catherine Mitchell from the Science Policy Research Unit and Eberhard Jochem from the Fraunhofer Institute for their patient persistence and encouragement.

Contents

1. Energy Policy in the EU - Diversity or Convergence?

Francis McGowan
School of European Studies
University of Sussex
Falmer, East Sussex
United Kingdom

1.1 Introduction

Over the course of the development of the European Union (EU), [1] there have been many attempts to formulate a common energy policy but with only limited success. The main reason for its failure has been the reluctance of member states to pool sovereignty in this highly sensitive policy area. Yet one might expect that a combination of factors would pull member states' energy policies in the same direction: the increasingly integrated European economy, the common challenges of the environment and international competition, even the increased activism of the Commission in areas which impinge upon the energy sector should be bringing about a convergence of national energy policies, creating the basis for a truly European energy policy. That this book focuses upon national energy policies might suggest that such convergence is not yet a reality. However, it is clear that the factors noted above have had an effect upon member states and that arguably that impact is becoming more significant. In this introductory chapter we summarise some of the developments at work in the countries examined in the book - France, Germany, Italy and the UK - as well as elsewhere in the EU, assessing the current balance between diversity and convergence. We review the development of European energy policy and consider its significance for national energy policies.

The orientation of official energy policies, and the balance of other policies affecting the energy sector, has shifted over the decades. For much of the post war period, there was an emphasis on fostering national energy resources and managing the transition to a more diverse energy balance, on the basis of a preoccupation with energy security (particularly since the oil crises of the 1970s). The other strand of

---

1 The contributions to this book refer to the EU generally except when discussing specific historical events, when the European Community or EC is referred to.

policy was also strategic, but in a broader sense, involving the use of the energy sector to fulfil wider economic objectives (the development of new technologies, control of balance of payments and inflation and the pursuit of social welfare). The energy supply industries were important mechanisms for pursuing these objectives and governments were able to exert influence through outright ownership or the allocation of special privileges within energy markets (such as the granting of exclusive rights or monopoly franchises).

More recently, however, the emphasis on supply security and strategic importance - along with the claim that energy is a "special case" - has been challenged as a result of radical changes in energy markets (in particular the fall in energy prices in the mid 1980s and the more favourable supply-demand balance in global energy resource markets), technological changes (the advent of information technology and the changing economics of various energy supply options) and broader political changes. As a result, other policy concerns such as the protection of the environment or the promotion of competition have begun to impinge upon the energy sector (McGowan, 1990). These changes have led to a shift in the interests involved in energy policy-making. In the past, the energy companies were the main targets and, arguably, the main beneficiaries of policy. As they were closely tied to government this outcome is perhaps not too surprising (though it would be wrong to say that policy was simply "captured" by producer interests). More recently, however, a much wider range of interests - with large consumers and environmentalists in the forefront - have sought to influence the priorities of energy policy.

Most governments have sought to maintain energy policy as a domestic responsibility, intervening either directly or through national firms to maintain some degree of sovereignty. Many regarded energy as being regarded as too important to leave to international market forces or to surrender too much to co-ordination within intergovernmental arrangements. Vulnerability to external energy shocks has been the norm for most countries (see Table 1.1). Whereas in 1960, the countries which now comprise the EU produced just under 70% of their needs, by 1970 that share had fallen to less than 40%. In the 1980s and 90s the share had risen to around 50% (largely due to increased energy efficiency and the development of domestic energy resources) though forecasts suggest dependence will increase in the future.

Table 1.1   European Energy Dependence, 1960-90 (m.tonnes oil equivalent)

| | 1960 | | | 1970 | | | 1980 | | | 1990 | | | 1993 | | |
	Energy Production	Energy Supply	Energy Self-Sufficiency	Energy Production	Energy Supply	Energy Self-Sufficiency	Energy Production	Energy Supply	Energy Self-Sufficiency	Energy Production	Energy Supply	Energy Self-Sufficiency	Energy Production	Energy Supply	Energy Self-Sufficiency
Austria	9.0	12.6	71%	8.1	18.3	44%	7.7	23.5	33%	8.8	26.1	34%	9.1	26.2	35%
Belgium	14	23.3	60%	7.1	40.3	18%	8	46.1	17%	12.6	48.3	26%	11.7	50.7	23%
Denmark	1	9	11%	0.3	20.2	1%	0.7	19.5	4%	9.9	18.3	54%	13.7	19.8	69%
Finland	6.0	10.2	60%	5.0	18.1	28%	6.9	25.0	28%	11.3	28.5	40%	29.3	47.1	62%
France	50.9	85.1	60%	42.6	147.3	29%	47	190.7	25%	104.8	221.2	47%	118.5	233.8	51%
Germany	127.1	144.5	88%	174.7	304.6	57%	184	359.2	51%	184.8	355.1	52%	149.4	337.2	44%
Greece	0.4	2.6	15%	1.7	8.1	21%	3.7	16	23%	8.8	22.1	40%	8.6	22.7	38%
Ireland	1.5	3.9	38%	1.4	6.3	22%	1.9	8.5	22%	3.4	10.6	32%	3.4	10.7	32%
Italy	21.8	48.1	45%	21.1	110.7	19%	20	139.2	14%	25.6	154.7	17%	28.6	156.5	18%
Luxembourg*	0	3.3	0%	0	4.2	0%	0	3.6	0%	0	3.6	0%	0	3.9	0%
Netherlands	11.2	22.5	50%	29.2	49.9	59%	71.8	65.5	110%	59.8	66.4	90%	67.9	69.7	97%
Portugal	1.7	3.4	50%	1.4	6	23%	1.5	10.3	15%	2.1	16.4	13%	1.9	17.6	11%
Spain	12	18.3	66%	9.7	38.4	25%	15.8	68.7	23%	31.2	88	35%	30.1	91.0	33%
Sweden	9.8	24.8	36%	6.5	38.0	17%	16.1	41.0	39%	29.8	47.8	62%	12.0	28.9	42%
UK	116.1	161.8	72%	101.8	207.7	49%	197.8	201.2	98%	207.5	211.8	98%	219.7	217.0	101%
European Union**	382.5	573.4	67%	410.6	1018.1	40%	582.9	1218.0	48%	700.4	1318.9	53%	703.9	1332.8	53%

Source: International Energy Agency

Note: The table shows the level energy production and of energy supply (ie energy used) in each member state. The percentage of energy production indicates the degree of self-sufficiency (or dependence) across the Community.

* - Luxembourg produces less than 0.1 m.tonnes oil equivalent

** This table covers those states which were members of the EC in 1995.

It has been difficult for countries to insulate themselves completely from developments within international energy markets, as the oil shocks demonstrated. Through a mix of subsidies, taxes and protectionist measures, governments have been able to maintain energy prices above, below, or in parity with "world prices" for traded energy commodities, often as part of efforts to support existing or emerging domestic energy supply options.[2] The issue of how far supply security can be equated with self sufficiency and how far countries need to be self sufficient in energy resources is highly debatable, but for countries with only a limited resource, decisions regarding the treatment of those resources (whether they be fossil fuels, nuclear power or renewable energies) will most likely be more sensitive than in a country with substantial and diverse resources.

Vulnerability, therefore, may help to explain the rather dirigiste approaches adopted in most member states over much of the last forty years. Of course, all the EC countries have market economies and all (or nearly all) claim that their policies towards the energy sector are based on the principles of the free market. However, notwithstanding the different ways in which "the free market" is interpreted in different countries (Shonfield, 1965; Albert, 1992; and Hart, 1992), the extent to which most countries' energy policies could be said to be operated on the same basis as most other sectors in the economy is highly questionable. Indeed, far from being run along the lines of the free market, the energy sector has been one of the few elements of a planned economy in many otherwise capitalist countries.[3]

The planning orientation is best demonstrated by the high priority given to forecasting as a tool of policy-making. While such techniques are common in many industrial policy concerns, the time horizons (often thirty to forty years) have been much longer than in other sectors, as is the degree to which they have been used as a guide to investment decisions (Baumgartner and Midttun, 1987). The "long termedness" of energy policy is also reflected in the chosen technologies, notably nuclear power, which are characterised by long lead times in construction and in operating life. The need to manage uncertainty in energy markets help to explain the

---

2 While it is of course debatable how far one can talk of a world price for energy products, the scale of the divergence in energy prices between member states reflects national preferences (in the form of covert subsidies or overt taxes) rather than market conditions. See European Commission, 1984.

3 The "distortions" in national policies are dealt with in the annual review of the International Energy Agency, various.

persistence of energy policy priorities and techniques throughout much of the post war period: institutional memory of market disruption and a perception that energy import dependence leaves economies exposed have been and remain powerful incentives for most governments in making energy policy.

1.2   Explaining Differences in National Energy Policies

If we focus on such aggregate factors, it is perhaps not surprising that we tend to see the similarities between national energy policies. The existence of rather similar communities of expertise in energy planning and policy-making might also prompt us to accentuate common features.[4] However, on closer inspection, we can observe important differences between states. To explain the differences in how European states have handled the more-or-less common agenda, we need to bear in mind institutional and cultural specificities of those states.

Prior to the 1980s, British energy policy conformed to the norm of a strong concern for supply security and for broader policy objectives. The government has had a direct influence upon the energy sector for many decades, dating back to the interwar period when it purchased a stake in what was to become British Petroleum and established a public corporation to manage the electricity grid and, more importantly, the post war period when the coal, electricity and gas industries were nationalised. The wider role of the energy in the British economy was a constant feature of government intervention (to meet the goals of reconstruction and the development of new technologies and resources).

Overall, the record of energy policy in the UK must be judged as mixed. By the beginning of the 1980s Britain was almost unique in Europe in enjoying self sufficiency in energy across a balanced portfolio of fuels, yet there were serious shortcomings in the performance of the energy industries themselves as well as in the conduct of government. Indeed, it could be argued that the problems of post war British energy policy could be summarised as government failing to intervene where it should have done (productivity) and interfering where it should not (technology

---

4   At least in the 1970s, the elite consensus on the priorities for energy policy and the techniques of analysis was such that one might be tempted to refer to an "epistemic community" of sorts (Haas, 1992). It is harder to identify such a community today, however

choice). Given this mixed record, it is not perhaps surprising that a radical government committed to rolling back the state should seek to redefine energy policy. In the early 1980s, the government signalled that in future market forces would be the main determinants of energy balances and that privatisation and liberalisation of the energy sector would be priorities (Department of Energy, 1982 and Lawson, 1992). While the policy has evolved slowly, by the early 1990s, British energy policy had been transformed. The major energy utilities were in private hands and were moving towards more competitive market structures.

In many respects, French energy policy developed along similar lines to the UK (long standing participation in the oil sector (Feigenbaum, 1986), widespread nationalisation after World War Two and the subsequent use of public firms for energy policy objectives). Responsibility for the energy industry has been shared by the Ministry of Industry, within which the Directorate for Energy resides, and the Ministry of Finance. The former has been responsible for the strategic development of the industry and the latter has been involved in questions of finance and investment and pricing. However, the nature of the French political system and particularly the close links between government and industry have led to a much more coherent pursuit of energy policy goals than in the UK. At times it has appeared that French energy industries have been better able to dominate thinking on key energy policy decisions than their British counterparts (Hayward, 1986; Lucas and Papaconstantinou, 1985).

Undoubtedly the massive nuclear programme of the 1970s and 80s demonstrates the coherence of French energy policy both in terms of the close links between government and industry and the consistency of policy over time. This programme was driven by the lack of alternative indigenous resources (aside from some gas and coal, the French energy resource base is confined to a declining coal industry and some small gas fields) and proved much more effective in delivering cheap electricity than any other nuclear programme either in Europe or the rest of the world (Thomas, 1987). The programme also illustrates the strength of the French energy industries, a strength which has more recently been demonstrated by the so far successful resistance of the French energy sector to pressures from those seeking to introduce greater competition or to encourage environmental protection.

At first glance the history of Italian energy policy appears to fit the pattern found in the UK and France, but in practice the gap between intentions and outcomes has been greater than in the rest of Europe. The state became a major participant in the energy sector during the interwar period but it was only after the war that this role was effectively exploited (in the development of the country's gas resources). This success proved rather unusual as the state oil and gas company and later the state electricity company (which was nationalised only in the 1960s as part of the "opening to the left" by the Christian Democrats) were used as instruments of patronage by all the political parties.

In theory policy was determined in the formulation of an overall National Energy Plan. However while this has been revised regularly over the decades, its forecasts and recommendations have rarely been realised. Instead the government has used the energy sector as a means of pursuing other policy goals (for example limiting price increases to control inflation). Implementing energy policy has also been rendered difficult by the planning process in Italy, one which allows local environmentalists to block new power station investments. The Italian environmental movement has been able to exploit these rules much more effectively than environmentalists in France or the UK, though it would be wrong to suggest that they manage to influence the overall balance of energy policy beyond this veto. Pressures to increase competition in energy markets have been historically weak but they are increasing as a result of the proposed privatisation of the energy industries.

In historical terms, German energy policy appears to be the exception to the tradition of centralised energy policies utilising publicly owned energy industries. The Federal Government no longer has any ownership stakes in the energy sector (the last were divested in the 1980s when it sold its stake in the industrial conglomerates VEBA and VIAG) while energy policy has explicitly celebrated the reliance on free market mechanisms (for example in the area of pricing). In practice, however, German energy policy has been in many ways as interventionist as any other in Europe, particularly in the last ten years with the rise of the environmental movement. Even the impression of separation between firm and state is undermined by the extensive ownership stakes of Laender and municipal authorities (Hardach, 1980).

For much of the post war period German energy policy was guided by a long-established consensus between government and the producer industries (and with them the engineering equipment sector). From the 1960s this was taken to involve a commitment to the then strategic technology of nuclear power (though the consensus was subsequently extended to coal through the "Contract of the Century", signed in the early 1980s, which provided support to the hitherto declining domestic industry). The policy was not without its opponents but it was not until the Chernobyl accident that the consensus broke down following the Social Democrats' decision to support a phasing out of nuclear power. Over the 1980s, moreover, the coal industry came under criticism not because of cost (even though by then it was three times the price of imported coal) but because of environmental problems, first acid rain and more recently the greenhouse effect (Boehmer-Christiansen and Skea, 1990). The debate on German energy policy is now characterised by real tensions between those who wish to take the new environmentalist agenda further and those who wish to restore the old consensus on nuclear power and coal (talks in 1993 designed to foster a new consensus proved unsuccessful). Attempts to liberalise German energy utility markets have met with only limited success.

Elsewhere in the Union, energy policies have followed two broadly similar patterns (though again we should be mindful of important differences amongst these countries). In the so-called cohesion states (the Mediterranean states and Ireland) the traditional energy policy goal of reducing reliance on imported energy has been matched by a commitment to the development of energy infrastructures: indeed, in recent years, concerns over supply security have diminished as energy prices have fallen. State monopolies have dominated the production and import of energy in most cases (Spain being an exception where private companies have played an important role in the electricity sector). In the small Northern European states (Belgium, Netherlands Denmark as well as the three new member states), supply security goals compete with the need to protect the environment in shaping policy priorities and technology choices. Most of these countries also have a more decentralised energy sector than the cases examined above (excepting Germany), with local governments often playing a significant role in energy supply and increasingly energy policy.

The differences between states are clearly demonstrated by the case of nuclear power, the most important of the long term technologies which dominated energy policy priority setting in the 1960s and 70s. At that time a European states have either invested in or committed themselves to develop the nuclear option initially on the grounds that it constituted the "fuel of the future", later on the basis that it provided a cheap and secure way of reducing energy dependence and most recently as an "environmentally friendly" source of energy. Yet the record of nuclear power has diverged quite widely over the years. France, and to some extent the UK, have been successful in overcoming opposition to nuclear power (as demonstrated by the construction of a new power station in the late 1980s) but only France has developed a full and successful commitment to the technology. Germany was able to develop an effective programme in the 1970s but widespread political opposition has foreclosed further development. Italy was not able to develop an effective programme due to a mix of policy failure and public opposition and was obliged to shut down existing capacity. Elsewhere, only Belgium, Sweden, Finland and Spain have significant programmes (with the future very much in doubt in Sweden and Spain). Elsewhere the technology has not been taken up either as a result of local opposition and/or changed perceptions of energy needs.

A rather different energy policy issue which all countries have had to address is that of environmental protection. The pervasiveness of environmental concerns - at the local, regional and global levels - renders it difficult to generalise about national tendencies to address the issue in energy policy. Nonetheless it is clear that the greatest efforts to develop "green" energy policies have been made in Germany, Scandinavia and the Netherlands. Specific environmental issues have affected national energy policies elsewhere: the siting of power stations has become extremely difficult in Italy thanks to a strong environmental movement. Moreover, green rhetoric has been much in evidence in most countries, though the degree of policy impact has varied considerably as the UK experience demonstrates. In the late 1980s the government "discovered" the environment as a political issue. However, while British environmental movements have been able to exploit that rhetoric as well as the turbulence in energy policy to promote some environmental goals, they have not been able to overcome the energy industries' successful opposition to major policy changes such as the carbon tax. French environmentalists, while at times politically influential, have not been able to refocus energy policy. In the cohesion countries environmental issues have had only marginal impact.

The other major policy development - privatisation and liberalisation - has only been embraced in a limited number of countries. While the UK is on the verge of privatising almost all of the energy sector, policies elsewhere have not been as ambitious in their scope or purpose. A number of countries - such as Austria, Germany, Portugal and Spain - have embarked on partial privatisations (whether at the local or national level) and in some cases - such as France - the programmes are rather limited. In some countries - notably Sweden - planned privatisations have been halted or slowed by political changes. The liberalisation of energy markets has also had a mixed record. The main liberalisation has been in the oil sector (where state trading monopolies have been abandoned in a number of countries over the last ten to fifteen years). In the energy utilities, a number of North European countries have reformed or are poised to reform their sectors but in other countries - notably France - there remains widespread opposition to introducing competition. In many of these cases, however, it may be the case that limited reforms may establish pressure for further changes in the future.

How do we explain these different approaches and outcomes? Without being determinist, it is clear that such basic factors as geography, geology and economic development have impinged upon national choices. In particular, energy policy decisions have been affected by resource endowments: the mix of energy resources has shaped the hierarchy of interests involved in energy policy-making and these in turn have influenced the priorities of energy policy. Beyond a consideration of producer interests, however, the balance of energy supply and demand has itself been influential. Most states have sought to exploit domestic energy resources (fossil or non-fossil), particularly in the wake of the 1970s energy crises, but there remain important differences in degree. There is no doubt that the UK's good fortune in possessing a well balanced portfolio of energy resources has affected its subsequent energy policy (particularly in the last ten years) while those which have been less well-endowed have adopted very different approaches to the management of those limited resources. Even amongst the latter, however, there has been an interesting difference between those which have endeavoured to maintain their indigenous energy industries (Germany, Spain) and those which have presided over - even hastened - their decline (France and Belgium, where nuclear has supplanted coal's role in power production and the UK, where government policies led to a radical contraction in the industry). The development of natural gas has been central to energy policy in the Netherlands and, to a more limited extent, in Denmark. Hydro electricity played an important part in energy policy-making in the past though its impact has diminished in recent decades because of the limited scope for new projects (only in Sweden is there the prospect of major developments and in that case new investments have been opposed on environmental grounds).

Political factors - in terms of institutions and ideologies - have also been influential. At the most basic structural level, the extent of centralisation in energy policy - and the energy options chosen - has generally reflected the overall balance of power between central and local government. Thus in France and the UK (where local government is traditionally weak) policy has been very heavily centralised whereas in the North European states a significant role has been played by generally more powerful local and regional branches of government, favouring in many cases less centralised energy options (such as CHP and energy efficiency). Beyond this territorial division of power, one can also observe the impact of policy styles and political traditions ("Colbertism" in France and political patronage in Italy being amongst the more obvious examples). The changing political shape of societies is also significant. Although there have been periods of consensus in most countries' energy policies, broader political changes have undoubtedly had their impact. The best examples of this impact are the rise of green movements in Northern Europe, and the British Conservative government's embrace of neoliberal economic policies. The impact of these political changes has been to alter significantly energy policy in some states more than others, indicating a dynamic between new and ideas and ideologies. Thus in Italy the effect of political upheavals provides some scope for new influences on energy policy whereas in France the relative stability of the political system and the cluster of interests involved in the energy sector has proven more resistant to change (at least so far). This interaction seems to favour the maintenance of differences between member states, slowing down the convergence of outcomes if not of issues.

1.3 The Changing International Setting and the Role of the EU

However, whether or not these new issues prompt energy policy convergence, it is clear that all countries have had to take them into account, even where their political importance or relevance is rather limited. One reason for this it that governments have had to react to policy proposals from EU institutions, proposals which reflect the broader influence of these policy ideas. Yet the role of European institutions has traditionally been rather limited in the energy domain. Indeed, international agencies in general have scarcely influenced national energy policies. The efforts of governments to maintain their sovereignty in energy policy matters have already been noted. That exercise of autonomy may be harder to sustain as the international context becomes more important for national energy policy making: energy markets and firms are becoming more global while many of the preoccupations of recent years such as trade liberalisation and environmental protection are inherently international. The issues which have arisen domestically, in other words, also appear on an international agenda.

The corollary to the defensiveness of member states' energy policy strategies has been a limited role for international co-ordination in the sector. The main institutional innovation was the creation of the International Energy Agency (comprising the OECD countries except - until 1990 - France) to handle supply disruptions in the oil market, to monitor developments in the energy sector and to review members' energy policies (Keohane, 1984). There has also been the development of a non-proliferation regime which has involved the management of trade in nuclear equipment and materials, though the effectiveness of the regime has been called into question on many occasions (Walker and Lonnroth, 1983). For the most part, energy has figured in the communiqués of western summits and UN debates but with little substantive effect.

The question of appropriate authority structures for energy policy however has been resurrected as the broader process of internationalisation in the world economy begins to impinge upon the structure of the energy industries.[5] The oil industry has, of course, been the quintessentially global industry, notwithstanding a wave of nationalisations from the 1950s to the 1970s (a wave which has since been reversed as foreign investment and joint ventures have blossomed). However, now other energy sectors are increasingly internationalised, a process encouraged by privatisation in both developed and developing countries.

These developments also reflect pressures on the demand side as the international markets for energy resources grow: coal is following oil as an increasingly specialised market with a limited range of countries exporting their lower cost resources (while many of their customers are scaling back their own higher cost

---

5 Some times this phenomenon is referred to as globalisation - for an account, see Ohmae, 1992. One attempt to deal with the concept from a political science perspective can be found in Grant, 1992.

industries). The gas and even the electricity markets have also become more open to trade. Moreover for some large users, energy price conditions have become an important criterion in corporate strategy. Such consumers of energy are likely to look increasingly beyond their national borders for supplies of energy and to question policies which restrict their ability to utilise the lowest cost source.

As a result of these changes, governments may find their autonomy in the energy sector is becoming even more limited, arguably as much as it is in other areas of the economy, such as steel, electronic components or finance. Moreover, these changes may be exacerbated by the changes under way within some countries. If the traditional mechanism of energy policy, the publicly owned energy corporation, is no longer to hand, governments may have to reconsider both their objectives for, and their approach towards, energy policy. Governments' scope for action may also be changing because of obligations to international agreements, particularly in the areas of trade liberalisation and environmental protection. Even if their own energy policies have not moved in the direction of greater competition/transparency and sensitivity to environmental impacts, commitments made in international fora may oblige them to do so.

It is clearly too early to judge how far these institutional developments will constrain national governments and regulate energy firms and markets. It could be argued that such changes may be necessary: if firms and markets are internationalising, should there not also be at least some element of international regulation? However, while the development of international authority structures might be thought desirable, the very real differences in interests and perceptions between governments are likely to limit the prospects for such initiatives. International governance might be considered an ideal accompaniment to these developments but, aside from some limited "regulatory" roles (opening up market access, attempting to limit emissions), it is likely to remain an ideal. In this context, regional policy solutions may be more realistic, obtaining what may no longer be possible at a domestic level. On the face of it, the EU is well suited to this task.

There is no doubt that energy should have played an important part in the Community's affairs: two of the three Treaties on which the EC is based are concerned with energy specifically: the ECSC and Euratom Treaties were devoted to coal (which then dominated energy balances) and nuclear power (which was seen as the future). A common market for other energy sectors was, by implication, addressed in the Rome Treaty. The gap between intentions expressed in the Treaties and the outcomes, however, has proved a large one for energy, and the Commission's attempts to secure the agreement of member states to a Community energy policy of any sort, let alone one reflecting the ideals of the treaties, were for many years unsuccessful (McGowan, 1994).

Throughout the 1950s and 1960s attempts to develop policies for energy industries were ignored or rejected by member states, who sought to retain control over the energy sector. The oil crisis of 1973/4 should have provided an opportunity for an EC role but instead proved to be a further instance of the Community failing to co-ordinate policy. Member states pursued their own policies or worked through the IEA (van der Linde and Lefeber, 1988). In the wake of these failures, the Commission attempted to develop a different approach to the management of energy supply and demand, involving the setting of target objectives (such as the reduction of energy imports as a proportion of total energy needs or the improvement in energy intensities). In these cases, however, the main concern was to change the structure of energy balances rather than the structure of energy markets (Commission of the European Communities, 1988a).

By the mid 1980s, therefore, the Commission had succeeded in establishing a place in energy policy making, but it was far from being central to member states' energy policy agendas, consisting instead of information gathering, target setting and most significantly R&D, activities which could hardly be described as constituting a comprehensive Community energy policy. From that time on, however, the scope for a wider, if unofficial, Community role in energy policy began to increase. This shift was partly a result of changes within the Community itself. The Single European Act marked the turn-around in the fortunes of the Community and was paralleled by renewed activism on the part of the Commission. That dynamism was most clearly seen in two areas - market liberalisation and environmental protection - which were in case impinging upon national policy agendas. These proved to be areas to which the policy techniques of the Commission and the competences and commitments of the relevant Directorates (DG4 and 11) were well suited. Thus the Commission was able to play an increasingly visible role in proposing policy and regulating the Community economy, including the energy sector.

There are other aspects to the Commission's energy policy agenda. It continues with its support for energy efficiency and renewables through research budgets (though this is only a small element) and other measures designed to encourage their use (such as recommendations for preferential terms for renewable sources of supply). It has developed policies for supporting the development of energy infrastructures, primarily in less developed areas of the Community. It has also sought to develop its responsibilities in the area of security of supply by seeking to join the International Energy Agency and play a more active role in crisis management.

Such a variety of activities would suggest that the Commission still anticipates a formal European competence in energy policy. Its attempts to formalise its role in the Treaty on European Union were unsuccessful: although the Commission was able to insert a relatively weak commitment to a Community role, in the very last stages of the negotiations a number of member states indicated their objections to it and obtained its removal in the final agreed text. The Commission was able to insert a commitment in the Treaty to reconsider the treatment of energy in the 1996 negotiations[6], however, and it is in the process of making the case for a Community energy policy role: at the beginning of 1995 it produced a green paper on energy policy and is putting together a fuller set of proposals addressing energy policy directly (Commission of the European Communities, 1995).

Whether or not the Commission is successful in securing a formal competence in energy policy, it is clear that it will continue to develop policies which act as constraints on how policy is pursued at a national level. The growth of this "regulatory" role is likely to remain the Commission's most potent means of influencing member states. The Commission is particularly well suited to operating in a regulatory mode (Bulmer 1994; Majone 1994). Moreover, most of the developments have been in the areas of competition and the environment, where a clearer European competence has emerged which effectively cuts across sectoral concerns. In these cases, it is possible to identify a more rounded perspective, based on broader principles (the benefits of competition or the need to protect the environment) than sectoral priorities and supported by a strong legal basis (indeed, the role of the Court in defining the scope of regulation has been of considerable importance) (McGowan and Seabright 1995).

---

6 See "Declaration on Civil Protection, Energy and Tourism" in the Treaty on European Union.

The emergence of the Commission's regulatory role on energy matters has reinforced its position in other respects. In particular, it has prompted many inside and outside the energy industries to take the Commission seriously. This is best illustrated by the refocusing of lobbying activities towards Brussels. The creation of specifically European associations for the electricity, gas, oil, renewables and conservation industries, the establishment of European branches in large energy firms' government affairs departments, and the increase in complaints and cases on energy matters taken to the Commission and the Court are symptomatic of the growing attempt to influence European institutions. Indeed, groups within member states are just as likely to try European conduits as a way of changing national policies (eg environmentalists' complaints to the Commission concerning government support for the nuclear industry in the UK).

Governments too can no longer ignore the Commission's activities in the energy sector. Whether it is the British seeking approval for privatisation plans, the Germans negotiating on Commission objections to coal subsidies or the French and Italians defending themselves against allegations of anticompetitive behaviour in the electricity sector, member state authorities must take into account the European dimension in almost all aspects of energy policy. The Commission's role in regulating for the environment is arguably less of a constraint upon national policy-making, if only because it has emerged on the basis of Council decisions, most of which require unanimity. In a sense, the member states have been careful to ensure that their room for manoeuvre on energy policy is not constrained by EU environmental regulation. When the Single Act was signed, member states attached a Declaration which stated that the new environmental rules would not affect member states' ability to develop their energy sectors. Similarly in the Maastricht Treaty, member states were able to secure unanimity as the method of decision making on environmental rules affecting the energy sector. Despite these limitations, however, there is still scope for the Commission to intervene regarding the way in which agreed decisions are implemented in member states.

The competition rules are a source of greater restriction as there is greater discretion allowed to the Commission in determining how far national conduct is in line with the Treaty. Increasingly the Commission is intervening on issues of market structure and conduct and the provision of government support in the energy sector. Some of these policy actions have a relatively long history (the attempts to curtail subsidies to the coal industry and the reform of oil monopolies). However, in the last few years the extent of Commission activities has increased markedly (a development not unconnected with its desire to develop the internal energy market). In instances such as British electricity privatisation, energy subsidies and others, the Commission has demonstrated a capacity and a willingness to intervene. However, the results of these interventions have been variable - in many cases it has been clear that the Commission has been under considerable political pressure and in some the Commission's bark has been worse than it's bite. Indeed, it would be wrong to suggest that the regulatory activities of the Commission are "unconstrained". There are clear limits to Commission interventions. At one level there is a question over whether the Commission should intervene (where the political calculation relates to the likely reaction of member states and whether or not Commission action would be seen as illegitimate). At another level, once the decision to act is taken there is the risk that its initiatives will be rejected in the Council or overturned by the Court.

1.4 Conclusion

Energy policy in Europe remains a national concern and important differences between national policies persist, despite the fact that most of the member states encounter broadly similar problems and challenges. The attempt by governments to maintain their independence in policy-making - while at odds with their own vulnerability in energy markets and with a more general commitment to integrate within the EU - has survived partly because of the very important interests at stake in this field and the power of the interests involved in the energy sector. Yet the strengthening role of EU institutions - the activism of the Commission and the broadening role of European law - can no longer be dismissed by member states: governments have to take more account of the European dimension to energy matters. However, even though the constraints imposed upon national policies are probably increasing in importance, they scarcely constitute a coherent EU energy policy. Each of the energy subsectors has been affected by European policies, but - beyond a set of promotional activities in such fields as information, research and regional planning - the most important interventions have been on the basis of broader policies such as the internal market and the application of competition rules on the one hand and environmental protection on the other. Thus a number of member states continue to resist the transfer of sovereignty on energy policy to the

18

Community yet they cannot ignore the EU as a constraint on their own autonomy. The rather confused division of labour which results however risks not only conflicts between member states and the Commission but conflicts between different strands of energy policy. It may be that the pressures of different policy objectives and the need both to reconcile these and to rationalise and regulate derogations from them, will push the EU towards a de facto energy policy, raising a number of questions about accountability and democracy. Ironically these problems will emerge because of member states' reluctance to consider and debate a coherent and common energy policy for the EU.

References

Albert, M. , Capitalism against Capitalism, London: Whurr, 1992.

Baumgartner, T. and Midttun, A. eds, The Politics of Energy Forecasting, Oxford: Clarendon, 1987.

Boehmer-Christiansen, S. and Skea, J., Acid Politics: Environmental and Energy Policies in Britain and Germany, London: Belhaven Press, 1990.

Bulmer, S., "The Governance of the European Union" Journal of Public Policy Vol 14, 1994.

Commission of the European Communities, The Main Findings of the Commission's Review of Member States' Energy Policies, Brussels: Commission of the European Communities, 1988a.

Commission of the European Communities, The Internal Energy Market, Brussels: Commission of the European Communities, 1988b.

Commission of the European Communities, The Application of the Community's Energy Pricing Principles in Member States, Brussels: Commission of the European Communities, 1984.

Commission of the European Communities, For a European Energy Policy, Brussels: Commission of the European Communities, 1995.

Department of Energy, "Speech on Energy Policy", Energy Paper 51, London: HMSO, 1982.

Department of the Environment, This Common Inheritance, London: HMSO, 1990.

Department of Trade and Industry, The Prospects for Coal - Conclusions of the Government's Coal Review, London: HMSO, 1993.

Feigenbaum, H., The politics of Public Enterprise: Oil and the French State Princeton: Princeton University Press, 1985.

Feigenbaum, H Samuels, R. and Weaver, 0., "Innovation, Coordination and Implementation in Energy Policy" in Weaver, R. and Rockman, B. (eds), Do Institutions Matter? Government Capabilities in the United States and Abroad, Washington: Brookings, 1993.

Grant, W., "Economic Globalisation, Stateless Firms and International Governance" University of Warwick Department of Politics and International Studies Working Paper 105, 1992.

Haas, P., "Introduction: Epistemic Communities and International Policy Coordination" International Organization, Vol 46, No1, 1992.

Hardach, K., The Political Economy of Germany in the Twentieth Century California: University of California Press, 1980.

Hart, J., Rival Capitalists, Ithaca: Cornell University Press, 1992.

Hayward, J., The Market Economy and the State, Brighton: Wheatsheaf, 1986.

Helm, D. Kay, J. and Thompson, D. (eds), The Market for Energy, Oxford: Clarendon, 1989.

International Energy Agency, Energy Policies and Programmes of IEA Countries, Paris: IEA/OECD, various.

Keohane, R., After Hegemony: Cooperation and Discord in the World Political Economy Princeton: Princeton University Press, 1984.

Lawson, N., The View from Number 11, London: Bantam, 1992.

Lucas, N. and Papaconstantinou, D., Western European Energy Policies : a Comparative Study of the Influence of Institutional Structure on Technical Change Oxford: Clarendon Press, 1985.

Majone, G., "The Rise of the Regulatory State in Europe" West European Politics, Vol 17 No 3, 1994.

McGowan, F., "Conflicting Objectives in European Energy Policy" in Crouch, C. and Marquand, D., (eds) The Politics of 1992: Beyond the Single Market, Oxford: Blackwell, 1993

McGowan, F., "Energy Policy" in El Agraa, A. (ed), The Economics of the European Community, Harvester, 1994.

McGowan, F. and Seabright, P., "Regulation in the European Community and its Impact on the UK" in Bishop, M., Kay, J. and Mayer, C., The Regulatory Challenge, Oxford: OUP, 1995.

Ohmae, K., The Borderless World, London: Fontana, 1992.

Shonfield, A., Modern Capitalism Oxford: OUP, 1965.

Surrey, A.J., "The Nationalised Energy Industries" in Gretton, J. and Harrison, A. (eds), Energy UK, Newbury: Policy Journals, 1986.

Thomas, S. D. The Realities of Nuclear Power, Cambridge: Cambridge UP, 1988.

van der Linde, J. and Lefeber, R., "IEA captures the Development of European Energy Law" Journal of World Trade, Vol 22 No5, 1988.

Vickers, J. and Yarrow, G., Privatisation - an Economic Analysis Cambridge: MIT, 1988.

Walker, W. and Lonnroth, M., Nuclear Power Struggles, London: Allen and Unwin, 1983.

2. French Energy Policy: the Effectiveness and Limitations of Colbertism

Dominique Finon
Institut d'Economie et de Politique de l'Energie (IEPE)
University of Social Sciences at Grenoble
Grenoble
France

2.1 Introduction

France is poor in energy resources: in 1992, it imported 96.3% of the oil consumed, 89.5% of the gas and 63% of the coal. Despite this, the French have been able to limit their reliance on foreign energy supplies to around 50%, due to the successful implementation of a nuclear power programme which produced 70 million tonnes of oil equivalent (mtoe) in 1991,[1] and also, in a less visible way, due to a policy of energy conservation which brought about a reduction of 35 mtoe over the period.[2]

The lack of energy resources provided the rationale for an active policy centred on the reduction of oil-dependence and the improvement of security of supply. Since it entailed the establishment of a new, highly capital intensive, technological system, this policy required both considerable time and stability to be successful. This stability was guaranteed by the centralised nature of the energy industries, their proximity to the State and the isolating of decision processes from the political arena. The nuclear programme thus become part of the great French tradition of "Colbertism", the tradition of strong state interventionism in industry and technology. Jean Baptiste Colbert (1619-83), the chief minister to Louis XIV, pursued effective policies of mercantilism, supporting manufacture and infrastructure construction (particularly with respect to roads, canals and the French Navy) for the affirmation of France's political power in Europe. Colbertism characterises public interventionism, notably in the industrial and technological fields.

---

1 The French official energy equivalence of the primary electricity is 0,222 MToe for 1 TWh.

2 The amount of 35 MToe is assessed by using a detailed bottom-up approach, which separates the effect of intersectorial changes of the GNP structure, from the unit consumption effect in the different sectors. This unit consumption effect is considered as energy savings. It is the results of behavioural and technical changes, linked directly or indirectly to the energy conservation policy.

However the French model is not without its limitations. Due to the lack of countervailing economic and political mechanisms, the obverse of the effectiveness of the implementation of policies has been a rigidity, an inability to adapt in the face of changes in the national or international context. The model is also less suited to decentralised and diversified areas of intervention, such as energy conservation and renewable energy sources, tending to encourage instead an over-concentration on certain, favoured energy technologies. But the model is now changing as the propensity to intervene weakens and the state institutions concerned face new policy challenges (such as the "internal market" and environmental protection).

2.2 French Energy Trends since 1973

Between 1973 and 1993, the French energy situation underwent some very substantial changes (see Table 2.1). Over this period, the national production of primary energy increased by 71.6 mtoe (from 41.8 mtoe to 113.4 mtoe), although the consumption of primary energy grew by the smaller amount of 37.2 mtoe (from 183 to 220.2 mtoe). The average elasticity of energy to GDP (Gross Domestic Product) was only about 0.4, and the decrease of energy intensity was 16% over the period. The ratio of energy independence improved substantially, moving from 22.5% to 51.8%.

Table 2.1: Changes in the Consumption of Primary Energy

(mtoe)	1973	1979	1985	1991	1994
Coal	27.8	31.9	24.1	19.1	14.1
Oil	126.6	118.8	84.3	91.3	93.5
Gas	23.3	21.0	23.3	26.1	29.6
Primary Electricity	13.3	25.1	58.0	73.9	85.8
Renewables[a]	2.0	3.0	3.9	4.2	4.1
Total	183.0	199.8	193.6	218.4	227.1

a: Excluding non-marketed wood materials (about 5 to 6 mtoe).

Source: Ministry of Industry (Observatoire de l'Energie), Les chiffres clés de l'énergie, Paris, Dunod, 1994.

This improvement was achieved by controlling consumption and by the vigorous development of nuclear energy: over the period, nuclear production increased from 14.8 TWh to 369.1 TWh, equivalent to 78% of total electricity production and 72.4% of total energy production. This development enabled 61.7 TWh of electricity to be exported in 1993. The growth in nuclear output made up for the decline in coal production, from 17.3 mtoe to 6.2 mtoe, and that of natural gas, from 6.2 mtoe to 2.8 mtoe. More effective exploitation of the hydro-electric potential (+ 2.1 mtoe) and of biomass resources (+ 2.2 mtoe) has also helped limit France's dependence on foreign energy sources. In fact, net imports of oil and oil products decreased from 124.4 mtoe to 85.8 mtoe, limiting the "oil dependence" to 38.5% of primary energy consumption, compared with 68% in 1973.

French sources of energy supply now appear to be better distributed, due to the increase of national supplies and the diversification of imported energy sources. Imports of natural gas, for example, have increased from 7 to 25.7 mtoe. Oil thus constituted only 40.4% of the primary energy consumed in 1993, nuclear and hydro-electric 38%, gas 13.2% and coal 6.3%. However, the successful development of nuclear power (through an extensive and well-resourced promotions campaign) reduced the potential for development of natural gas, by effectively limiting the latter's outlets for its electricity production and domestic space heating. On the demand side, final energy consumption between 1973 and 1987 experienced both years of stagnation and of decreases as well as years of slow growth (less than 2% per year). In total, final demand has increased by only 8.3% over the whole period, whereas the economy as a whole grew by 30%.

Table 2.2: Changes in GDP, Energy Consumption and Dependency, 1987 - 1991

(%)	1988	1990	1991	1992	1993	1994
GDP	4.2	2.4	0.6	1.2	-1.5	2.6
Primary energy	3.1	2.3	1.6	1.5	-0.5	0.8
Final energy	4.2	2.6	1.5	2.9	-0.2	
Ratio of Independence	48.3	47.8	47.8	49.3	50.6	51.6

Source: Ministry of Industry (Observatoire de l'Energie)

Between 1988 and 1993, energy efficiency has scarcely improved, though the ratio of energy independence has stabilised (see Table 2.2). The commissioning of nuclear power stations has slowed down, whereas the growth of consumption has revived again because of the belated recovery of economic growth and a substantial drop in oil prices. By 1990, the elasticity of energy consumption to GDP had moved close to 1, as it was before 1973. Subsequently, however, and in marked contrast to previous recessions, the downturn of the 1990s was not accompanied by a major reduction in consumption.

2.3   The Importance of the State

The Scope of the Public Sector in the French Energy Industries

The effectiveness of French energy policy is explained in large part by the power of the state's means of intervention. In West Europe, France is one of the countries where the dominance of the public authorities is most marked. This dominance takes the form of the existence of a powerful regulatory capacity, the presence of public or semi-public enterprises and the existence of government agencies in all the energy fields.

Oil

In 1928 the French government passed a law granting itself a monopoly in the importation of oil and oil products: the state then contracted out import quotas to the subsidiaries of the multi-national oil companies and to French companies. By this means the government has been able to create a petrol refining industry in France prior to the second world war. It has also systematically implemented a policy of control over French oil supplies, by allocating an increasing share to French companies (Despraires 1969). In contrast to other market economies, the French state regarded dependence on the international oil oligopoly as placing national sovereignty in question. Although the government never sought to exclude them from the domestic market, it nonetheless privileged the French firms and fostered their development.

The Compagnie Française des Pétroles (CFP), created in 1924 as a joint state-private company (the State currently holds 35% of the capital), was for a long time given preference over the multi-national companies, basically Exxon, Shell and BP (Catta, 1990). Centring its activities on Iraq, it was not interested in exploration in the French colonies. After the war, the state created some small public enterprises for this purpose, later attempting to develop a more effective instrument for the exploration and production of crude oil, and, after 1956, for marketing Algerian oil. Following a series of mergers between these companies and other private small refiners and distributors (Caltex, La Mure, etc), between 1960 and 1966, an integrated company, Elf-ERAP, was created. This was destined to become the "secular arm of the state, its sword in the world oil struggles, alongside the CFP which serves as the shield" (Pean and Serenti 1982).

Elf-ERAP progressively expanded its market share to 25% in the 1970s. It merged in 1970 with ANTAR and in 1975 with the Société Nationale des Pétroles d'Aquitaine (SNPA), a mixed company, which operated the natural gas deposit of Lacq, in order to benefit from the economic rent of gas to reinvest outside Algeria. The objective set jointly for Elf-ERAP and CFP was the control of the production of an amount of oil at least equivalent to France's annual consumption. This objective would have been achieved if OPEC had not put an end to oil concessions.

Electricity

The Nationalisation Act of April 8, 1946 allowed only a little scope for an independent electricity sector outside self production, a municipal distribution sector (Picard, Beltran and Bungener 1985) and production by nationalised firms in the coal, railways and steel industries.[3] In creating EDF (Electricité de France), the Act sought to ensure that the electricity industry would be planned and coordinated, and to permit equitable supply conditions to all users throughout the national territory. Only the city-owned enterprises such as Bordeaux, Grenoble, Metz and Strasbourg (which currently distribute about 6% of low voltage electricity) were not nationalised. EDF has a monopoly over transmission and imports and exports, as

---

3 CDF (the coal company) produces 10TWh, the SNCF (railways) 1,1 TWh, the steel companies 2,0 TWh. There is also the exception of the Compagnie Nationale du Rhône. It owns the different Rhône dams built for the "canalisation" of the river. It produces 14,5 TWh in 1991.

well as being the exclusive owner of the distribution concessions. Although subject to the regulatory control of the Ministry of Finance and the Ministry of Industry, particularly with respect to annual investments, financing and tariff increases, EDF has enjoyed strategic and managerial autonomy since 1984. Regulation takes the form of an incentive-based contractual relationship with the government. Under this regime, a Planning Contract is drawn up every four years, containing various objectives: the lowering of prices in real terms, the reduction of indebtedness and an improvement of production and marketing performance. The most recent contract was signed in January 1993.

The French state also has a strong presence in the nuclear industry, both in the building of reactors and in the nuclear fuel cycle. The Commissariat à l'Energie Atomique (CEA) controls, through its holding company CEA-Industrie, the entire capital of COGEMA which has a monopoly over all aspects of the fuel cycle, from the extraction of uranium to reprocessing. CEA also owns a large part of the capital of Framatome (36%), the builder of nuclear reactors. The remainder of Framatome's capital is shared, on the State's side, between EDF with 10% and Crédit Lyonnais with 5%, and, on the private side, Alcatel-Alsthom (formerly CGE) with 46%; the staff of the company own 3%. Since 1986, CGE has sought to take control of Framatome to integrate it with its activities of electric construction. But the CEA has succeeded in maintaining public control over the nuclear sphere. In reality, however, CEA-Industrie has little control either over COGEMA or Framatome. [4]

Natural Gas

In 1946, the (then-manufactured) gas industry was subject to the same logic of nationalisation as electricity. The initial objective was to ensure that the industry, which had been stagnant since the start of the century, was technically updated. Gaz de France (GDF) is institutionally less powerful than EDF, due to the influence of the private oil interests upstream and the limited significance of the gas sector for the economy (Beltran and Williot 1992). Moreover the distribution activities of EDF and GDF for power and for gas are jointly carried out, a situation which has largely restricted competition between the two fuels, perpetuating GDF's inferior status.

---

4 On the relationship between EDF and Framatome, see Thomas, 1988 and Finon, 1989

GDF had neither a monopoly in production, nor, following an amendment in 1949 of the Nationalisation Act, in transmission. The French oil companies wanted to participate in the marketing of gas from the South-West of France, notably the Lacq deposit, which was exploited after 1957. In that region, natural gas was transported and resold by the Société Nationale de Gaz du Sud Ouest (SNGSO), 70% of which was controlled by Elf-Aquitaine (GDF holds the remaining share). In central France a joint company owned by GDF (50%), Elf-Aquitaine (40%) and also Total (10%), the Compagnie Française de Méthane, held the transport concession. The public establishment had a monopoly over imports and exports, and also over distribution concessions, notably in the SNGSO area, apart from a small municipal distribution sector which was retained by the Nationalisation Act and comprising 16 "régies" (local public monopolies) providing 5% of residential gas sales.

Since 1965, GDF has been able to conduct an active policy of development and of diversification of its gas imports, sharing these between the Netherlands, Algeria, the former USSR and Norway: in 1991 Russia supplied 34.5%, Algeria 31%, Norway 20% and Netherlands 15.3%. But its market position has been weaker than that of EDF, since it has few captive customers. The level of supervision by the Ministry has also been tighter. Control over prices was more severe (for anti-inflation reasons), causing operating deficits over many years and it was not until February 1991 that the state established a contractual, incentive-based regulation with GDF broadly equivalent to EDF's "Contrat de Plan" with the signing of a three year "Contract of Objectives". This contract contains objectives of profitability, of reducing the company's indebtedness, of improvement of the productivity (outside import costs), and of commercial innovations (Revue de l'Energie, 1991).

Coal

The state has also been dominant in the coal industry ever since the nationalisation of the coal mines created Charbonnages de France (CDF) in 1946. From being at the centre of energy policy after the war (producing 60 m.tonnes in 1958), the French coal industry has experienced a continuous decline since the early sixties (see Table 2.3). The first rationalisation program (le Plan Jeanneney), decided in 1960 and implemented in spite of a long strike in 1963, aimed to cut production by 6 m.tonnes by 1965. A second programme (Plan Bettencourt) was agreed in 1968 with the aim of reducing output by a further 26 m.tonnes by 1975, 50% of production. Under pressure from large users, notably EDF, the government has chosen not to protect the price of French coal - the priority being low prices for all fuels on the domestic market - and has instead financed the operating deficit of the company.[5]

The state has also sought to control coal imports by granting a monopoly over foreign coal purchases in 1948 to a state purchasing agency, the ATIC (Association Technique des Importations Charbonnières). The intention was to ensure the security of supplies and to protect domestic production, while, at the same time, giving French consumers a stronger negotiating position. ATIC has operated as a compulsory intermediary for the various purchasers, allowing them to benefit from its negotiating power. Because of internal market obligations, this monopoly was suppressed in 1990, and, henceforth, it has contractualised its relationships with importers-resalers and with consumers.

Table 2.3: **Trends in French Coal Supply 1954 - 1974**

(m.tonnes)	1954	1960	1964	1970	1974
Production	56.3	58.3	55.8	40.6	26.8
Imports	10.9	12.0	18.5	16.6	20.1

Source: Ministry of Industry (Observatoire de l'Energie).

Energy Conservation

The national energy conservation policy is led by the ADEME (Agence de l'Environnement et la Maîtrise de l'Energie), under the authority of three ministries, Industry, Research and Environment. Formally, the country's energy conservation strategy was led between 1974 and 1982 by the Agence pour les Economies d'Energie (Energy Savings Agency), and between 1982 and 1991 by the Agence Française pour la Maîtrise de l'Energie (AFME), also responsible for renewable sources of energy.

---

5 The direct subsidy for the mining deficit amounted to 2,8 thousand million Francs in 1991, which was 35% of the value of production (8 billion Francs). It was 3,2 billion in 1987. The total subsidy to CDF includes also compensation for social charges for retired miners, and for the financial charges of the regional reconversion investment. The total amount was about 6 billions in 1991 (it was 6,8 billions in 1987).

The Organisation of Energy Policy

The Ministry of Industry is responsible for the formation and implementation of energy policy.[6] This Ministry shares the supervision of the energy companies with the Ministry of Finance. However, the shaping of energy policy is strongly influenced by the exercise of energy planning carried out every five years within the scope of the formation of the five year economic plan under the aegis of the Commissariat Général du Plan. A mechanism for indicative planning until the 1960s, and an overall forecasting process since then, the plan formulates targets by mutual consultation between the ministries, the companies, the unions and the consumers (MacArthur and Scott 1970; Shonfield 1976). This consultation process enables the long term plans of the main actors in the economy to be coordinated within the framework of more general criteria formulated by the government such as limiting "dependency", reducing supply costs, and subsidising investment costs in certain industries.

Since 1973, the four successive plans which have paralleled the progress of energy policy (7th plan in 1975, 8th plan in 1980, 9th plan in 1983, 10th plan in 1990) have varied in their importance. The first two were constrained by the decisions concerning the development of nuclear power - decisions which were taken directly by the government - while the formulation of the eighth plan in the new political context of 1983 was an opportunity to recognise nuclear overcapacity and to formulate the necessary adjustments. In 1990, on the other hand, the Tenth plan was explicitly formulated as a simple forecasting exercise, locating the uncertainties in the economy and exploring various policy options.

Energy policy is implemented through the investment programmes of the public enterprises, regulations (such as control of oil imports), price and tax controls over various energy products, public subsidies, controls over the public enterprises (investment authorisations, financing, prices etc.), and reorganisation of the energy

---

6 The Direction Générale de l'Energie et des Matières Premières (DGEMP: Energy and Raw Materials General Authority) consists of the Direction des Hydrocarbures (DHYCA: Hydrocarbons Authority) and the Direction du Gaz de l'Electricité et du Charbon (DIGEC: Authority for Gas Produced from Electricity and Coal), and the Service d'Utilisation Rationnelle de l'Energie (SERURE: Department for Rational Energy Use) in the field of energy.

industries. The policy is set within the framework of constraints imposed by the Ministry of Finance in regard to taxation, prices and state investments. Some of these constraints are linked to European commitments, notably in the fiscal field (principally VAT and excise duty). For the most part, energy prices have been liberalised in line with a more general removal of price controls. The price of heavy fuels and naphthas was set free in July 1978; the other oil prices in January 1985, that of coal and gas for industry since December 1986. Only the tariffs of electricity and some prices of gas for residential and tertiary sector remain controlled.

The Political Realities

Decision-making on energy policy is not confined to the guidelines made or formulated in the Plan, however. Many major decisions are taken within the framework of the close relations existing between the state energy enterprises and the ministries, and these restrict the scope of the planning process (Martin 1974; Saumon and Puiseux, 1977; Lucas 1977). This has particularly been so in the case of French oil policy (the accelerated import penetration of oil products and the creation of a second state oil company between 1960 and 1967), of the rationalisation of the coal industry (the Jeanneney Plan in 1960 and the Bettencourt Plan in 1968), and of nuclear policies (with an important role played by the PEON Commission).[7] Most of the decisions have been made behind closed doors, away from the relatively open procedures of the Plan and away from the political arena.

Politicians therefore play a low-key role in the matter of energy policy. There has only rarely been a Minister of Energy in the government (from 1981-1986 and in 1992). This self-effacement is explained by the fact that higher levels of power seem to hold similar views on the main lines of energy policy, combining national interest with economic modernisation (French independence with respect to American oil or nuclear interests, technological prestige, industrial redeployment, reduction of the costs of oil, etc). The effectiveness of the implementation of the successive policies has perpetuated the legitimacy of this decision-making group, which is very largely

---

7 The Commission pour la Production d'Electricité d'origine Nucléaire (PEON: The Commission for Nuclear Energy Production) was disbanded following the political changeover in 1981. PEON brought together the leaders of the industrial groups concerned with nuclear production - EDF and the CEA - and the representatives of the two Ministries concerned, Industry and Finance. See Simmonot, 1978.

autonomous and not open to the initiative and control of politicians. The latter have only been obliged to intervene when there were conflicts between state enterprises. This autonomy of energy policy can be seen clearly from the very limited nature of the changes in policy following the political changes in 1981 and 1986, despite the new policies loudly proclaimed at the time of assumption of power by the new parties.

This autonomy is also partly explained by the weakness of parliamentary power in France, under the constitution of the Fifth Republic. The French parliament has little control over the specialised areas of industrial policies. Energy policy has only been debated on rare occasions and has never been the subject of substantial research, unlike in other industrialised democracies. There was an exception in September 1981 when, after a change of government, a temporary commission examined the options available in energy policy and proposed innovatory policies in energy conservation, some of which were brought into effect. There were two other parliamentary debates in the 1980s: in 1986, after the Chernobyl accident and in December 1989. A degree of progress was achieved in 1984 with the creation of the Parliamentary Office of the Evaluation of Technological Choices which, occasionally, deals in some detail with topics related to energy problems (the safety of nuclear reactors, waste storage, prevention of the greenhouse effect); but its influence has remained limited. The centralisation of French institutions also contributes to the autonomy of energy policy-making. In contrast to more decentralised countries, French regions and local bodies have limited powers, and have little voice in the implementation of industrial policies.

The dominance of the public powers and the autonomy of energy policy is explained, finally, by the particular sociology of the French state, which is characterised by the presence of the powerful "state corps of engineers" (Corps des Mines, Corps des Ponts) which monopolise the management levels of the Ministries and of state enterprises (Hayward, 1986). The proximity, cohesion and osmosis between the civil service and the state enterprise manage is one reason for the effectiveness for the effectiveness of the large state- or joint-owned enterprises. This was the case with the recovery of coal production and the implementation of the hydro-electricity programme in the 40s and 50s, with the setting up of a state oil industry in the 60s, and with the nuclear programme in the 70s and 80s. The corps of engineers in general were dominant in the areas where large technical projects that were "in the national interest" have to be implemented (Bauer and Cohen 1986; Cohen 1991). The legitimacy which they derived from their image of competence and disinterested service has enabled them to protect these projects from the uncertain effects of political interference, in the name of economic rationality and of the national interest, with which they strongly identify themselves. They thus have succeeded in imposing their viewpoint on the policy process, and have avoided any serious confrontation of expertise. There is a danger, however, that the "national interest", thus defined, may easily correspond to the organisational interests of the state institutions and enterprises.

In such institutional circumstances, the "errors" which result from the absence of effective economic and political controls may be hidden. There is in effect an inversion of the relationship between the energy enterprises and the supervising administrations, due to the osmosis of the top leaders and officials and to the imbalance of economic expertise between the two groups. In this context, and leaving aside the technical success of the programmes, insufficient attention was paid to domestic and international market trends. As a result the French energy system experienced excess production capacities, rapid obsolescence and high social costs (Martin 1990). The instance of the nuclear overcapacity of the 1980s was preceded by the accelerated replacement of coal by hydro-carbons during the 1960s in the name of the search for competitiveness, and the excessive revival of coal production between 1950 and 1955 (Toromanoff 1969).

The state has effectively monopolised the overall capital and technological resources for the energy sector, even though the opportunity cost of this monopolising has never been evaluated. In 1980, for example, the energy sector absorbed 64% of industrial investments stricto sensu, when French industry was experiencing difficulties in redeploying to the areas of high technology manufactured products (Martin, Criqui and Finon 1984).

The free market critiques of the 1980s had little effect on the organisation of the energy industries and little effect in redefining the role of the State. With the change in government in 1986 to 1988, the Chirac cabinet, elected on the basis of a free market programme, failed to implement substantial reforms in this area, even when the Ministry of Industry was headed by an ultra-liberal predisposed to denationalise EDF, GDF, COGEMA and Elf-Aquitaine and to abolishing the import monopolies of coal and oil. In fact the government never considered the privatisation of any of these enterprises. It only agreed to decrease slightly the state's share in the capital of Elf-Aquitaine, to abandon the system of oil import control (without abrogating the law of 1928) and to adjust the supervisory mechanisms dealing with the state energy enterprises, granting them greater managerial autonomy.

It has only been with the development of the "internal energy market" debate at the European level that there have been moves to modify the internal rules of the game (as will be seen below). Without bringing the basic institutional organisation into question, the appearance on the scene of a new power centre (the European institutions) advocating new rules is progressively moving things in a liberal direction, less oligopolistic and more decentralised.

2.4   Energy Policy in the 1970s and 1980s

After the first oil crisis, France adopted the same type of policies as other industrialised countries dependent upon oil imports: acceleration of nuclear investment, diversification of imports, promotion of national energy sources and energy conservation. Thanks to its centralised political system, however, France distinguished itself from these other countries by its success in implementing its nuclear programme. But the closed-off nature of its decision-making processes certainly did not help it to adapt to the economic changes of the late 1970s. As a result the entire French energy system has had to adjust to nuclear overcapacity.

Attempts to Reduce Oil Dependence (1973-1983)

In 1974, the level and structure of the energy forecast for 1985, set in 1970 by the Sixth Plan, were totally revised. The forecast for total primary consumption was reduced from 284 to 240 mtoe, while the share of oil was reduced from 63% to 40%, that of nuclear electricity was increased from 14% to 25%, that of natural gas from 12% to 15% and that of coal from 7% to 12%. The oil importers were to ensure that no single country was to supply more than 15% of energy consumption. Bolstered by Gaullist nationalism, France sought to go it alone, refusing to become a member of the International Energy Agency so as not to appear to give allegiance to the USA. To provide secure oil supplies, France banked on bilateral arrangements with producer countries, making contracts on a state-to-state basis with some of them (Saudi Arabia, Iran, Iraq, Mexico, etc). The French companies were encouraged to diversify their production activities to North Sea and African or Latin American countries. Oil prospecting was revived in France through a system of tax incentives. Moreover, international market oil price increases had immediate repercussions on the price of fuel and oil products within France (despite concerns about inflation), thus encouraging rapid adaptations of behaviour.

The focus of domestic energy policy was placed on the nuclear programme, which in France did not encounter the same regulatory and political obstacles as elsewhere. Six orders for nuclear power stations were placed per year in 1974 and seven in 1975. Orders followed thereafter at a rate of five or six per year until 1982. In five contracts made between 1974 and 1980, EDF ordered 32 reactors of 900 MW and 16 reactors of 1300 MW. Six others, of which four were 1300-MW and two were 1450-MW, were then ordered between 1983 and 1986. In comparison with countries confronted with planning difficulties, the success of this crash programme was very evident (see Table 15): in 1985 34, 000 MW were in operation; the capacity of 45, 000 MW, which was the initial target, was attained in 1987.

The scale of the programme allowed for the rapid diffusion of learning effects, reinforced by market structure (a single buyer, a single builder, a single technique) and the possibilities of standardisation and of industrial planning permitted by the stable regulatory framework (one which was enhanced by the close proximity of the nuclear safety authorities and the state promoters of the programme) (Thomas, 1988; Martin, Criqui and Finon 1984). Effective control over time-schedules and costs, and the relatively satisfactory performance of the reactors once in operation reinforced the apparent success of the programme. While it is true that, compared with the 1974 estimates, the stated unit cost of investments increased by 100% in real terms, international comparisons (OECD, UNIPEDE) show that French reactors were 80% cheaper than German or Japanese ones. The counterpart of this cost control by standardisation, however, was the regular appearance of generic faults in nuclear plant, which on each occasion occurred to as many as a dozen or two dozen reactors, requiring costly repairs. [8]

---

[8] For the 900 MW and 1300 MW series of reactors, corrosion problems appeared in 1988 on the steam generator tubes and on the pressuriser tubes, due to design faults: a replacement programme of steam generators is under way on 24 reactors (cost per reactor: 600m. Francs). At the end of 1992, cracks appeared on the covers of 13 reactors

The priority accorded to the nuclear programme naturally worked against the attempt to establish energy conservation measures and to promote renewable sources of energy. The budget of the two small agencies created in 1974 and 1975, Agence pour les Economies d'Energie (AEE), and Délégation aux Energies nouvelles, grew slowly (from 152 m. Francs in 1975 to 683m. Francs in 1980), remaining much lower than the 5b. Francs of the non-military budget of the Commissariat à l'Energie Atomique. However, the country's interventionist tradition and fear of energy dependence have allowed a certain stability in policy and have permitted some institutional learning in even these new fields of intervention.

The first measures taken in 1974-1975 were mainly either educational, intended to change behaviour in the use of heating (limiting temperatures, reducing the annual heating season, etc), or were of a regulatory type intended to improve the quality of new housing (insulation standards and technical performance criteria). Starting in 1975, the government also gave individuals the inducement to deduct expenditures on energy savings from their taxable income. Later, in 1979, after an increase in the budget of AEE, subsidies of 400 Francs per toe saved were granted in all sectors. The investments induced by these measures grew to the estimated figure of 14 to 15 billion Francs per year in 1980, of which two thirds were in the residential and service sector. Various activities promoting the renewables (aid to research and development, support for investment, etc) were also implemented to promote geothermal heating (in late 1974), solar energy (late 1975), heat recovery (by a law voted through in 1977 which reshaped the legal framework for the development of district heating systems) and of biomass and synthetic fuels (January 1981).

Table 2.4: Gap Between Targets Set in 1974 for 1985 and Implementation

(GW)	France	Germany[a]	Italy	Japan	UK	USA
Targets	45/56	44	26.4	49	15	205/260
Implementation	34	16	1.3	23	8.7	70

a: West Germany

Source: IAEA, OECD - Uranium: Production, Demand and Resources - Paris, OECD, 1975

of 900 MW and 2 reactors of 1300 MW - these will have to be replaced (cost per reactor: 400m. Francs).

2.5   The 1980s: the Adjustment to Nuclear Overcapacity

The same political and institutional factors which furthered the technical success of the French nuclear programme also delayed its adjustment to the slow-down in the rate of electricity demand growth. The closed nature of the decision-making process did not take account of unofficial forecasts which anticipated overcapacity if the rate of orders was maintained (Chasseriaux, Chateau and Lapillone, 1979). The choice of this level of orders assumed a rate of growth in electricity demand of 6 to 7% per year, whilst overall economic growth stagnated at a level of 2 to 3% per year. Moreover, the change of government in 1981 did not immediately bring about a significant shift in the nuclear programme, despite the original plans of the socialists.

The socialist government's change of policy was due on the one hand to a desire for economic credibility in the eyes of the business community and on the other to an acceptance of high growth scenarios (5%) despite the world recession. On the basis of these forecasts of high growth the new government had also decided to boost domestic coal production (changing the 1990 target from 13 to 20 mtoe) and to reinforce the policy of energy conservation measures. These steps were designed to satisfy various components of its electorate, notably ecologists disappointed by the continuance of the nuclear policy. The Agence Française de la Maîtrise de l'Energie (French Agency for Energy Conservation) (AFME) was created in 1982. Its budget was supported in particular by the allocation of a fund for economic recovery, the Fonds Special Grands Travaux (FSGT or Special Large Works Fund), which was given 1.6 b.Francs for the AFME. These allocations were intended to fund district heating systems, the insulation of public buildings and public housing and, in certain cases, investments in energy-saving industrial equipment. In order to better detect possibilities of saving energy and to promote local initiatives, AFME placed offices in each region.

It was not until 1983, during the preparation of the Ninth Plan, that the problem of overcapacity was finally recognised. Whereas investment decisions had been taken on the basis of forecast requirements of 430-450 TWh in 1990, the new forecasts anticipated levels of 322-346 TWh, on the basis of a growth rate of 1.2 to 2%. (Commissariat General du Plan 1983). The government decided to reduce the rate of investment in nuclear reactors, though without abruptly cutting off orders as private companies would have done: the objective was to ensure the survival of the nuclear construction industry (see Table 2.5).

The Decline in Domestic Coal Production

The nuclear overcapacity required an accelerated phasing out of the conventional power stations (over 6,000 MW were put in reserve between 1983 and 1987) and reduced recourse to recently-built coal-fired stations (the five blocks of 600 MW put in service between 1981 and 1985). The policy of attempting to maintain domestic production of coal, followed since 1975 and reinforced in 1981, was abandoned: the target for 1990 was reduced from 20 mtoe to 10-12 at first, and then to 8 mtoe in the restructuring plan of 1986.

Table 2.5:   Reduction in the Rate of Orders Placed for Nuclear Reactors

1987	1982	1983	1984	1985	1986
0	3	2	2	1	1

Source: Author's calculations

The contract between CDF and EDF did not impose major burdens on the utility by comparison with Germany or the UK, either with respect to prices (which were aligned with international prices) or of tonnage commitments (2 m.tonnes per year between 1987 and 1992 for example). Moreover, since there is no EDF contractual commitment for buying CDF's power production, these sales are presently strongly dependent of the performance of the nuclear reactors. The Government only maintained financial support to CDF, by the mechanism of a permanent subsidy to compensate its deficit and the costs of reconversion. CDF, therefore, proceeded to a voluntary policy of pit closure, reducing manpower from 51,000 in 1980 to 26,000 in 1989 and 19,600 in 1991. Production declined from 20 m.tonnes in 1979 to 10.4 in 1993 whilst average productivity grew significantly (the productivity of the miner grew by 92% between 1985 and 1991 in Lorraine). Average coal extraction costs decreased in parallel, from 800 to 505 F/t (francs 1991) from 1985 to 1991 (Commissariat General du Plan 1991). This policy was accompanied by measures aimed at retraining miners and at minimising the regional impact of mine closures: CDF created an industrialisation fund to encourage the growth of new enterprises.

The Intensive Promotion of Electricity Use

In 1983, the government encouraged EDF to follow an active policy of developing exports, with the target of selling 50 TWh in 1990. At first, it had to overcome EDF's reluctance to look abroad for outlets, because of the utility's public service culture of basing itself primarily on the needs of French users.[9] At the same time, the government granted EDF all its support in developing a very ambitious marketing policy within France, to win 50 TWh of extra outlets between 1983 and 1990, compared with the predicted trend, of which over 30 TWh was to go to industry, 11 TWh to the residential sector and 9 TWh to the service sector. The government authorized EDF to give direct financial assistance to promote new equipment to both industrial users and the manufacturers of electrical equipment, the total annual amount of this assistance being approximately 1 billion Francs.

In industry, EDF has striven to accelerate the development of electricity uses by the promotion of all the existing techniques: induction, conduction, arc, resistance, radiant heating for drying, heat pumps, mechanical recompression of steam, variable speed motors. The company has also encouraged bi-energy solutions by very attractive seasonal tariffs. EDF has, in addition, promoted the development of modern, competitive electrical substitutes for traditional processes: production of hot air, steam production (in summer), etc. The implementation of this strategy, however, has run into difficulties, since the gap between the price of a therm of electricity and one of fuel has increased since 1985 due to the fall in oil prices. The expansion of electricity outlets has been easier to achieve in the residential and service sectors, not so much because a therm of electricity is less expensive there (even though the tariff structure is, in fact, more favourable for residential use than for industrial use), but because the reduced cost of investing in electric heating has been very attractive to house builders. Domestic electric heating is therefore very widespread in France: since 1983 it has been chosen for at least two thirds of new accommodation.

---

[9] In 1991, exports were 52 TWh. In 1993, the level reached 61.7 TWh, of which Switzerland took 18.5, Italy 15.2, UK 17, West Germany 3.5, Benelux 5.9 and Spain 1.6 TWh.

The overall result of this policy has been relatively satisfactory in terms of meeting its original objectives. Starting from a level of 261.4 TWh in 1982, the total consumption of electricity rose to 349.6 TWh in 1990, an annual rate of 1.75% p.a., compared with an average rate of economic growth of 2.3%. It should also noted that this market penetration by electricity was made at the expense of potential outlets for gas, particularly in the housing sector. Gas output only grew by 4 mtoe between 1980 and 1990, against 7 mtoe in Germany and 10 mtoe in Italy, over the same period. Gaz de France therefore found itself after 1985 with a surplus of one or two billions cubic metres, in spite of the flexibility of the "take or pay" clauses of its gas contracts.

Fluctuating Energy Conservation Policies

At the same time as its promotion of electricity use, the government also pursued energy conservation measures. In the 9th Plan of 1983, the target for savings to be achieved, between 1982 and 1990, was 40 mtoe. No doubt the conflict between these objectives was tempered by the fact that EDF promoted essentially the high performance uses of electricity. However, its policy of promoting electric heating aroused regular criticisms from AFME and from the Ministry of Industry (CGEMP 1988). Nonetheless, EDF maintained its policies while AFME's policies of financial incentives were radically altered: having reached the record level of 4.4 b.Francs in 1984, the energy conservation budget (of AFME and FSGT) decreased rapidly, particularly after the change of government in 1986.

Table 2.6: Trends in Energy Conservation Expenditure

(b.Francs)	1983	1984	1985	1986	1987	1988	1989	1990
AFME budget[a]	1.50	2.37	1.44	0.74	0.44	0.42	0.42	0.45
Other Credits (FSGT)	-	2.00	1.38	1.45	-	-	-	-

a: includes allocations of Special Large Works Funds (FSGT) managed directly by AFME
Source : Ministry of Industry, (Observatoire de l'Energie), Chiffres clés sur l'Energie), 1992.

The fall in energy prices reinforced the new government's relatively more free market approach. The new government did not seek to compensate for the effect of this price fall by raising taxes on energy. The government tried to do away with AFME, but in the face of opposition from regional authorities, the government limited itself to sharply reducing its budget, to 0.44 b.Francs. The return to power of the Socialist Party, in 1988, did not immediately change this direction of policy, although in 1989 a report requested by the Prime Minister concluded in favour of promoting energy conservation measures (Brana 1989). Other objectives were considered to be more important:: the fight against inflation, to achieve economic growth and the harmonisation of European taxes (an objective which would involve a fall rather than an increase in the average level of French taxes on energy).

The growth of environmental concerns (regarding acid rain and the greenhouse effect) led to a revival of energy conservation policies at the beginning of the 1990s. The government's wish to link up energy conservation and the protection of the environment came into effect in 1991 by the merger of AFME with two environmental protection agencies.

On balance, the effects of AFME's incentive-based policies were positive. These policies contributed to the re-shaping of energy consumption in France, although without achieving the savings target laid down by the 9th Plan (40 mtoe between 1982 and 1990), due to the reduced efforts after 1985. The Ministry of Industry and ADEME estimate that the amount of energy saved from 1973 to 1990 was 35 mtoe. The policies AFME implemented in association with other departments (housing, transport) obtained notable results in various areas (Radanne and Puiseux 1989). In new housing, energy consumption per housing units was almost halved between 1970 and 1980, due to improvements in construction standards. In old housing and service sector premises (hospitals, schools, public buildings), 7 million "heating diagnoses" were realised, and followed up by the implementation of insulation measures or of changes in the means of heating in numerous cases. A number of heating networks were built, now serving 6 million "housing unit equivalents", based on a wide variety of energy sources: boilers with traditional fuels; geo-thermal deposits (in the Paris region); industrial thermal waste and the incineration of household waste. Energy intensive industries (iron and steel, chemicals, cement works, glass works, paper mills) have invested in heat-retaining materials, condensers, and the improvement of heat exchangers. In transport, thanks to notable improvements in performance of vehicles, the consumption of new cars has fallen by 2 litres/100 km during the course of the 1980s.

Apart from its activities directed at the rational use of energy, AFME has strengthened its research work on renewable sources of power, particularly on solar heating, the conversion of light to electricity, windmill-type power generators and on the obtaining of methane gas from waste and from wood. Some success has been obtained, but the fall in the price of oil has put a stop to many projects.

2.6   Current Trends in French Energy Policy

Due to the persistence of low prices and abundant supplies in energy markets and the surplus of electricity produced by nuclear power, energy policy has not been a priority for French governments since the middle of the 1980s. However, since 1986, energy has retained some political importance for three reasons: the uncertainty of the world oil market (rekindled during the Gulf Crisis), the increase in environmental concerns (with respect to local, regional and global pollution), and the achievement of the Internal Market, which requires a certain degree of change to be imposed on the organisation of the energy industries.

Moreover, while energy issues have not been very high on the political agenda, the current government has sought to maintain some legitimacy in this field. In November 1993, Parliament - which is traditionally rarely consulted on energy policy - debated the deregulation of the energy networks. Moreover, an open debate was held on energy and environmental policies from June to October 1994, involving a wide range of interests (energy companies, trade unions, consumers, local government, NGOs, ecologists, and independent experts) beyond the traditional closed and technocratic decision-making process. After extensive regional consultations and six national colloquiums on major issues (technological risks and environmental control, CO_2 mitigation policy, transportation policy, the role of regional and municipal government, etc.) the Souviron Report published in December 1994 challenged the legitimacy of past policies, particularly the emphasis given to nuclear power at the expense of energy conservation and renewables, including the non-transparency of nuclear costs assessment. Different measures have been proposed for opening up the decision-making process and for promoting more intensively energy conservation, renewables and less centralized options of supply by the creation of new incentives (Souviron 1994).

It could be that some practical steps could follow in the light of these recommendations, given that the Souviron Report reflects a certain change of the balance of concerns regarding energy and environment policies, and the diminution of the public energy companies' influence upon the Administration. However it is hard to consider these developments as constituting a radical change because institutional structures have not been challenged nor have certain key issues, including European liberalisation and the future of the nuclear industry (reprocessing vs opened nuclear fuel cycle) been fully debated.

Official Energy Prospects for the Year 2005

It is hardly necessary to underline the uncertainties affecting the variables which impact on energy supply and demand: the economic growth rate, the price of oil and other imported forms of energy, international environmental protection regulations, etc. Coping with these uncertainties, the report of the Energie 2010 group, prepared under the auspices of the Commissariat Général du Plan in 1991, set out the following forecasts for energy balances. The forecasts scarcely affect the main priorities of French energy policy as it has developed since 1973.

Since it is difficult to forecast for several years ahead, it is necessary to construct scenarios with a view to highlight the various possible trends (see Table 2.7). The range of the forecasts for 2005 is wide, between 225.5 mtoe and 280.3 mtoe. Scenario A groups together the main factors tending to slow growth in energy consumption: slow economic growth (1.6% per year), high oil prices ($35 a barrel in 2005 at constant prices) and a vigorous energy conservation policy. Scenario B, on the other hand, brings together those conditions likely to produce a high growth in energy consumption: sustained economic growth (3.6% per year), moderate oil prices ($21 in 2005) and a weak energy conservation effort. The final consumption which results from Scenario A falls within a trend which is a continuation of that of the period 1973-1985: an annual growth of electricity consumption below 0.5% stimulated primarily by road transport and, to a lesser extent, by the residential-services sector. Scenario B predicts an annual growth in final consumption of nearly 2%, brought about in particular by a strong upturn in industry's energy consumption.

Starting from these two basic scenarios, another hypothesis (Scenario C), which adds a vigorous energy conservation policy to Scenario B, is examined. This predicts that energy savings of 28 mtoe between 1990 and 2005 are possible, as opposed to 12 mtoe in the unmodified Scenario B (for identical levels of economic growth).

Table 2.7:   Past and Future Energy Balances

(mtoe)	1973	1990	1993	2005A	2005B	2005C
Final Consumption	152.4	179.4	187.6	187.7	236.7	220.0
- Iron and steel	14.1	8.3	7.2	6.4	8.5	7.9
- Industry	44.2	44.6	44.7	44.0	64.4	58.1
- Residential-services	58.6	77.9	84.5	83.6	95.8	89.8
- Agriculture	3.1	3.3	3.4	3.0	4.0	3.5
- Transport	35.4	45.3	47.8	50.7	64.0	60.7
- Own use and losses	19.7	22.2	19.8	25.0	27.5	26.3
- Non energy uses	11.0	12.2	12.8	12.8	16.1	15.8
Primary Consumption	183.0	213.8	220.2	225.5	280.3	262.1
- Coal	27.8	19.0	14.0	17.4	21.6	21.3
- Oil	126.6	90.6	89.0	82.0	108.7	96.6
- Natural gas	13.3	26.3	29.2	28.7	40.0	34, 9
- Hydro-electricity	10.7	14.2	15.2	16.3	16.5	16.5
- Nuclear power	3.3	69.6	81.2	90.7	108.8	107.4
- Renewable energy	2.0	4.2	4.2	4.8	5.1	5.2
- Traded Electricity	-0.7	-10.1	-13.6	-14.4	-20.0	-20.0
Electricity (TWh)						
- Net production	174.5	399.5	448.5	502.6	620.6	
- Domestic consumption	171.3	349.0	386.7	437.6	530.6	
- Export less imports	3.0	45.6	61.8	65.0	90.0	

Sources: Observatoire de l'Energie and Commissariat Général du Plan, Energie 2010.

France's level of oil independence will stabilise at around 50%. Oil imports could only be less than this in the case of slow economic growth (when there would be a decline from 90.6 mtoe in 1990 to 82 mtoe in 2005). In the case of high economic growth, oil imports will increase by 6 mtoe assuming a strengthened energy conservation policy, or by 18 mtoe without such a vigorous policy. The commitment to the nuclear option will be maintained. Nuclear production will increase from 69.6 mtoe to 90.7 mtoe in Scenario A, and to 108 mtoe in the two other scenarios. Nuclear energy is the only national energy source capable of substantially contributing to the satisfaction of growing requirements, in view of the limitations of exploitable hydro-electric potential, the exhaustion of the Lacq deposit and the progressive decline in French coal production (to 3.6 mtoe in 2005). The promotion of the use of electricity will probably be continued, with consumption increasing at an average annual rate of 1.5% between 1990 and 2005 in the case of slow economic growth (1.6%) or of 2.8% for high growth (3.5%). The share of electric heating in new housing should remain at around 65% of the market. The growth in electricity requirements and of long-term export sales will probably lead to the disappearance of electricity overcapacity around 1998.

The above scenario was perceived until the mid of 1994 as implying the need for the revival of the nuclear program at a rate of about one 1450 MW reactor per year. The output of the new reactors during their first ten or fifteen years will be used for medium- and long-term export contracts: the contracts signed in 1990 and 1991 imply exports of about 70 TWh in 2000 (corresponding to the output of 8 1300 MW reactors at a load factor of 75%). Without any real political debate, the choice had been made to construct reactors for export purposes, but without exceeding a ratio of exports/production of 15%. However, in June 1994, EDF and the Ministry of Industry decided to delay any new order of nuclear plants to 2000, because of the slower growth of electricity demand than forecasted and the development of a bulk of cogeneration units which have reduced EdF's prospects for increased demand.[10]

---

[10] However, other numerous entries into independent production (730 MW) using motor diesel were attracted in 1993-94 by the high level of buy-back rates in superpeak hours. The Electricity Nationalisation Act of 1946, as amended in 1955, obliges EDF to buy back electricity from independent producers (up to 8 MW). This obligation was temporarily suspended at the end of 1994 (except for electricity produced from cogeneration or renewables).

It is not expected to utilise combined cycle power stations for base load requirements. This form of power is thought to be only competitive with the nuclear power stations if used for less than 3500 hours annually, and is considered to be subject to too many uncertainties regarding the price of gas (DIGEC, 1993). A combined-cycle power station of 650 MW may be built for experimental purposes, however, since the authorities wish to gain experience in this technology. Moreover, the growing influence of the ecology groups could lead to a genuine diversification of power supplies, particularly for the base load. Indeed, the gas requirements will grow only slightly between 1990 and 2005, unless there is strong economic growth: between 8.6 and 13.7 extra mtoe of gas will be imported in 2005 under Scenarios B and C, against 2.4 mtoe in Scenario A. The additional output of gas for use for electricity production will only amount to between 1 to 2 billion cubic metres.

Seeking Flexibility in Energy Production

The strength of the state energy institutions has provided the stability required for an energy policy based on long-term considerations. Significant results have been obtained in reducing oil imports and in increasing energy effectiveness. But the high degree of specialisation of the French energy system, increasingly dominated by electricity, has as its counterpart a degree of vulnerability, particularly with regard to the possible shift in public opinion against nuclear technology. Given the problems associated to the ageing of the nuclear power installations and the risks of nuclear-powered electricity, the nuclear regulatory authority has sought to win its autonomy from the institutions promoting the technology. Since the Chernobyl incident in 1986, there has been an increase in the severity of controls and precautionary checks, resulting for example in the replacement of the steam generators of all 900-MW reactors. Maintaining social acceptability also requires considerable effort to find a solution to the problem of how to store long-life nuclear waste (in contrast to their attitude to the siting of reactors, local governments have shown strong hostility towards the suggested sites for nuclear waste storage). [11]

The uncertainties with respect to the growth of national energy requirements and with respect to future trends on the international market also provide good reason

---

11 In December 1991, the French Parliament voted for a law setting up tests for the "packaging" and storage of nuclear waste for 15 years, before the building of the final geological storage centres.

for an increase in the flexibility of the system. But any substantial widening of the range of available technical options for energy production would require adjustments in a number of institutional provisions and regulations. The conduct of the state energy monopolies, particularly pricing policies which disregard the effects of geography on costs of supply, have prevented the emergence of niches in the energy market where various new technologies could be tested or where a more general diffusion of new techniques could be undertaken. To allow other techniques of electricity production to develop would require the lowering of institutional barriers: for example, by changing the Nationalisation Act, applying new rules for the setting of resale tariffs for electricity by EDF and by organising "competitive bidding" through a regulatory body. In the same way, the geographical differentiation of electricity prices would encourage the use of renewable sources of energy in the countryside. Similarly, in the towns, EDF's monopoly of distribution effectively prevents local bodies from joining together to develop cogeneration in urban energy systems (for urban heating, electric power, etc.). More generally, the culture of growth within EDF has rendered it less concerned about the rationale to promote electricity savings, arguing that this should be pursued through price signals.

With respect to gas, the legal monopoly that GDF has for each new gas concession prevents the expansion of local distribution grids in areas where GDF's required financial rate of return (12% in real terms) is not achieved.[12] Such hurdles have combined with other factors (such as the subordination of GDF to the commercial objectives of EDF due to the mixing of gas and electricity distribution and the higher costs of heating appliances) to limit the role of gas in competition against other fuels.

The ministries, conscious of these institutional limitations, are seeking to change them without putting at risk the centralized nature and the public status of the enterprises. They have adopted a more favourable attitude towards demand-side management, thereby preserving their legitimacy (by showing goodwill) and keeping control over policies in this field. Several similar initiatives have been taken in this direction since 1992 - mainly in the period when there was for a short time a Ministry of Energy in the government (Energie Plus 1992) - including moderate

---

12 Currently gas distribution reaches 5,000 communes, 60% of the population. In 1991, GDF managed to prevent a reform in the Nationalization Act which would have allowed communes not at present supplied with gas to contract for distribution concessions with other enterprises.

corrections of the tariffs for the resale of electricity by EDF (1992) and a contract between EDF and ADEME relating to a commitment to cooperate on demand side management trials and the promotion of renewable energy sources in several rural départements (February 1993). In this agreement, EDF commits itself to invest 100m. Francs per year in such policies during the years 1993-1996. During the Balladur government (March 1993-May 1995), a new impetus has been given to such policies, with the Decembner 1994 decision to use 200m. Francs per year of the Rural Electrification Funds (3b. Francs per year) to finance DSM and renewables.

In addition, certain renewable energy sources have benefited from developments within agricultural policy. In 1992 ADEME was assigned the objective of promoting the consumption of wood by developing, with the help of regional bodies, outlets in public buildings and collective housing. The purpose is to increase the consumption of wood (marketed and non-marketed) from 9 to 13 mtoe between 1990 and 2005. Also, in the face of the problem of the reduction of cultivated farm acreage (under the Community Agricultural Policy), the government decided in 1993 to promote the use of bio-fuels, primarily from rapeseed, on set-aside land. Governmental encouragement takes the form of a special subsidy and tax break.

The Growing Acknowledgement of Environmental Constraints

Whilst the environment has never been a priority issue for the French political class, the rise in environmental awareness after 1985 has been reflected in increased government attention to the issue. The Ministry of the Environment has consolidated its influence since 1986, relying in particular on Community environmental policy. In the area of atmospheric pollution, the European Directive of November 1988 on emissions from large fixed installations became a decree in June 1990. The regulations oblige existing installations to reduce discharges of SO_2 and NOX (40% in 1998 for NOX) and set rules on emissions for new plants (90% reduction in SO_2 emissions for plants over 500 MW). In addition, a tax on SO_2 and NOX emissions (150 Francs for SO_2) was created in July 1985 and extended to other emissions in May 1990. The revenue from this tax is used to subsidise the improvement of industrial equipment. European influence has also played a key role in tightening the standards on lead content in motor fuels (from 0.4 to 0.15 g per litre in 1991) and the fitting of catalytic converters to all new cars, despite pressure from French manufacturers of small cylinder vehicles. Significant tax relief on unleaded petrol has led to its rapid spread since 1989 (30% of fuel sales by 1991).

The effort to impede the greenhouse effect has likewise been the subject of significant political-administrative mobilisation since 1989. Following the report by the Inter-Ministerial Group on the Greenhouse Effect, which examined the various economic feasibilities of prevention (GIES 1990), the government set itself the target of limiting CO_2 emissions to the stabilised level of 2 tonnes per capita in 2000 (instead of 1.8 tonnes in 1990). There is relatively little room for manoeuvre for France in comparison with other countries due to its success in implementing its nuclear programme and its energy-efficiency profile. Only a doubling in the rates of energy conservation would make it possible to achieve such a level. The French position in the international negotiations is therefore difficult; in spite of the consistency of its energy policy, which has made possible a significant reduction in atmospheric emissions, including CO_2 (see Table 2.8), the negotiations only refer to future reductions in emissions, without taking past efforts into account.

Table 2.8:   Trends in French Atmospheric Emissions

	1980	1988	1990	1992
SO2 (kt)	3339	1227	1188	1206
(of which electricity)	(1224)	(233)	(321)	(330)
NOX (kt)	1646	1448	1489	1519
(of which elec.)	(287)	(81)	(105)	(111)
CO2 (Mt)	503	369	381	388
(of which elec.)	(111)	(35.3)	(43)	(44)

Sources: CITEPA. Documentary Research.-Paris, 1991; Ministry of Industry (Observatoire de l'Energie). Chiffres clés sur l'énergie. Paris, 1993.

No specific programme has been set out to do this. The French government accepts the principle of a high tax on fossil fuels on condition that all the industrialised countries enforce such a tax. The amount of the tax has not been specified, although the inter-Ministerial group has mooted the level of 1000 Francs/tonne of carbon. The conditionality of this measure avoids the government having to confront the energy intensive industries hostile to this tax. Moreover government awareness of the global environment issue has not prompted to revive its energy conservation policy, with, even after, the creation of the ADEME in 1991. Since this date, the ADEME's budget on energy efficiency and renewables has been maintained at a very low level of 0.45b.Francs, falling to a new low of 0.31b. in 1994.

2.7   The Impact of the Internal Market

European integration necessarily entails a weakening of the French State's instruments of intervention, given the emphasis on the free market and the transfer of responsibilities to the Community level. The institutional interplay is tending to open up energy policy to other participants, as a result of the adjustments in French regulations to Community principles and the greater potential for recourse to the European authorities. Moreover, the nationalist culture is already tending to mellow under the influence of European "acculturation". In 1989, the French government authorised an alliance between the French and German nuclear industries for exporting reactors.[13] Still more revealing, in 1992, France joined the International Energy Agency, choosing to play its cards in a multilateral game, facing the oil producing countries with a view to developing stability in the energy situation.

The construction of a greater Europe is in conflict with the French interventionist tradition, public service culture and the interests of public monopolies. It has in effect necessitated the suppression of State aid and the abandoning of commercial oil and coal import monopolies. It may yet require the dilution of various GDF and EDF monopolies, with, in particular, the introduction of third party access to the grid. The adaptation to the Commission's wishes has been made with a greater or lesser resistance, subject to the negative effects expected in terms of restrictions on long-term efficiency and security of supply. Enacting the adaptations has thus been easy with respect to State aid,[14] transparency of prices, tax harmonisation (VAT, excise duties) and the system of oil imports, whilst marked resistance has been shown in respect of the electricity and gas monopolies.

---

13 Framatome and Siemens have created a joint subsidiary, Nuclear Power International, which markets the reactors outside Germany and France.

14 Since 1990, the State no longer guarantees EDF borrowings. In addition, EDF gives a 5% rate of return on its share capital whatever its results.

Reforming the French Oil Regime

It will be remembered that the 1928 Act granted the oil and refined products import monopoly to the State, which then delegated it to the oil companies in the form of import authorisations [1]. This system had often sustained attacks from Community officials since before the Single Act. From 1979, after negotiations with the Commission, the government agreed to give up product quotas. But the general system, was maintained until 1992 due to resistance from the public technostructure (the Corps des Mines) because it was considered as a tool without equal for organising the handling of stocks during a crisis. The easing of the 1928 Act nevertheless enabled genuine competition to be exercised with the independent importers-suppliers (especially supermarkets), which have won 30% of the motor fuel market. The completion of the internal market made it inevitable that the texts would come to conform with the facts. The new Act at the end of 1992 only maintains a few obligations connected with security of supply (constitution of strategic stocks, obligation to notify the authorities in order to monitor market development, partial obligation to transport under a French flag), even if the oil industry considers these obligations unnecessarily restrictive.

In parallel with this change, the State has conceded increased strategic autonomy to the French oil industry. Elf-Aquitaine diversified mainly into chemicals and pharmaceuticals and established itself in the USA. The evolution of the ideological and political context has led since 1986 to the progressive reduction of the state's share of capital in the two enterprises. The last and very symbolic decision has been the privatisation of Elf-Aquitaine in February 1994.

Table 2.9: Evolution of State Shareholdings in French Oil Companies

	1975	1986-88	1992	1994
CFP-Total	35%	31.7%	5.4%	5.4%
Elf-Aquitaine	66%	53.9%	51.6%	13.0%

Source: Annual Reports, various.

---

1 This system was burdened by other obligations: strategic stock-piling, sailing under a French flag, obligation to buy French oil, the system being supplemented by the general verification of prices, abolished in 1978 for fuel oil and 1985 for motor fuels.

Defending the Electric and Gas Monopolies

In 1988-1989, at the start of the Community debates on liberalising access to the electricity and gas grid, the Ministry of Industry adopted a favourable position on third party access (TPA). In its opinion, economic liberalisation could be used to serve the aim of nuclear electricity exports. It did not perceive the risk that the organisation of the French electricity industry might be brought into question, even though its foundations were one of the principle factors in controlling nuclear power costs and French competitive advantage. For reasons of consistency, the gas sector must undergo the same type of reform as the electricity sector.

After a period of reflection, EDF from 1990 adopted a hostile stance to TPA, but a favourable one to the free transit of electricity. It favoured cooperation rather than competition with the other vertical electricity companies, estimating that cooperation better serves its aim of utilising its surplus production capacities and of long-term exports associated with the constructing of dedicated power stations. It also perceived the risk of its structures being destabilised, with a separation of the distribution and production-transport segments.

The common culture of the French energy "technostructure" helped the Ministry of Industry to align its position with that of EDF and GDF in 1991. Since then, it has built up several lines of defence with them, in particular the firm refusal to abolish the two public companies' import and export monopolies in the face of administrative injunction from the Commission in July 1993. Such an abolition would in no way weaken EDF's internal position. GDF, on the other hand, is another matter. It would need to agree to SNEA becoming an importer-transporter, and give up its plan to recover SNGSO when the Lacq deposit is exhausted. But the Ministry of Industry would quite like to see the emergence of a second leading player in the gas sector

However the threat of European economic liberalisation has not been without its effect upon the internal efforts of EDF and GDF to improve productivity and redefine their relations with the local authorities. In order to protect itself from free market criticism, EDF adopted a type of incentive-based management with decentralisation to "profit centres" (or "results centres"). Moreover, whereas the system of distribution franchises lost any sense between 1946 and 1990, EDF and

GDF, encouraged by the Ministry of Industry, wanted to anticipate the risk of local communities claiming control of their grids under European influence. The content of the concessions was thus redefined in 1991-1992: they will enable the local authorities to be more actively involved in some of the facets of distribution (environment, quality of service, etc). Nevertheless, the Ministry did not want to amend the Nationalisation Act to enable them to take back the distribution concession for their own account or to entrust them to mixed ownership companies.

Moreover some French players would be ready for a number of concessions, in particular the abolition of the legal barriers to independent electricity production, and to monopolies of import and exports. A governmental commission of high level civil servants was mandated in July 1994 to define the French position which could be seen as a compromise by the European Commission. The "Mandil report" published in January 1994 proposed reforms which could be seen as important progress, given the rigidities within the energy utilities (DGEMP 1994) These reforms, which are strongly criticised by the trade unions in spite of their limitations, would have to be concretised in 1995 after the presidential election. But it preserved in fact the essential principles of the organisation of power and gas industries : the national monopoly of transport, distribution and marketing with the central coordination by EDF and GDF. Over and above the freedom of imports and exports but under authorisation, the Mandil report proposed:

- the suppression of legal barriers on independent power production, with the adoption of the competitive bidding procedure, the organisation of long term contracts with EDF, and the unbundling of accounts;
- the introduction of very limited third party access. For gas, it would be opened to very big importers or consumers clearly identified (the Elf-Aquitaine's south-western regional distribute, subsidiary ammonia producers, etc) in the framework of long term transactions. For electricity, the access to the network would be opened only for export licensed producers and also between self-producers and their industrial installations. Independent transmission lines could be also allowed to be built. But municipal distributors would be excluded from this proposal of limited provision of TPA ;
- the establishment of a small regulatory authority which would manage and control these new procedures.

The Mandil report also supported the desire of local authorities to be more involved in the control of the distribution, but in a limited way. It proposes only to opening the gas distribution concessions up to competition in areas not currently supplied, even if GDF is eventually interested.

The Mandil report's propositions have been formalised in the concept of the Single Buyer which was proposed by the French government in autumn 1994 as an alternative to the principle of negotiated Third Party Access. The other line of defence involves intense lobbying around the issue of public service obligation and an extensive interpretation of the concept of general economic interest which could permit the maintenance of monopoly rights (under article 90.3 of the Treaty of Rome).

The public structures in the electricity and gas sectors retain strong legitimacy. It is very symptomatic that the Balladur government named in March 1993 has not sought to realise its initial plan to apply the new European regulatory proposals on competition to industries of a monopolistic nature in the energy industries and open up the capital of the public energy monopolies to private shareholders.(Le Monde 10th February 1993). By contrast he adopted a much more reformist position for the telecommunications for instance, envisaging a partial privatisation of France Telecom and accepting in June 1993 the European Directive on telecommunications deregulation. At the same time, Chirac's presidential program maintained a strong defence of the public utilities in the face of pressure from the European Commission.

2.8   Conclusion

The development of the French energy sector since 1973 confirms the capacity of the French State to implement its energy policy effectively. This effectiveness is based on the substantial scope of the public sector devoted to aspects of energy, working through a powerful state-industrial complex in a way that is greatly insulated from the political arena. This effectiveness applies both on the supply side, with the successful implementation of a centralised, technological system that is capable of reducing oil dependence, and also on the demand side, i.e. energy conservation. The efficiency benefits are less obvious since centralised, interventionist practices are not well adapted to the demand side, characterised as it is by a multiplicity of actions to be implemented, by the need to mobilise a large number of players and by a low level of political visibility.

The organisation of energy policy, however, sustained by a dominant culture which is somewhat "supply-oriented", interventionist and nationalist demonstrates a number of adverse characteristics: over-investment, the seeking for technical virtuosity, the capture of political power and of the regulator, the imbalance of information (expertise), etc. Lacking the competition of other economic and political forces, "technocratic" effectiveness is a different matter from allocative efficiency, but this is hidden, to a degree, behind closed-off decision-making mechanisms.

In France, interventionism in the energy sphere remains an intangible principle, the legitimacy of which has not really been touched by free market criticism. On the other hand, the move towards a Greater Europe and the pursuit of decentralisation will both work to moderate this interventionism, to mobilise a less simplistic version of nationalism, to channel it within the limits allowed by the affirmation of new powers. It is probably through such institutional developments, which permit the confrontation of divergent cultures and political groups, that a degree of rethinking may occur in the future. Even though limited, this change in the configuration of powers should permit a certain degree of technological diversification to emerge, guaranteeing more flexibility in the face of the uncertain future, more political legitimacy, and probably a greater internalisation of the environmental costs associated with energy production and use.

References

Bauer, M and Cohen, E Les grandes manoeuvres industrielles, Paris, Belfond, 1988
Beltran, A and Williot, J Le Noir et le Bleu: 40 ans d'histoire de Gaz de France, Paris, Belfond, 1992.
Catta, E Victor de Metz: de la CFP au groupe Total, Paris, Total Edition Press, 1990.
Cohen, E Le Colbertisme High Tech (Colbertism in the High Tech Age), Paris, Seuil, 1991.

Commissariat Général du Plan - Energy 2010, Report of the Group chaired by M. Pecqueur, Paris, La Documentation Française, 1991.

Commissariat Général du Plan - Rapport du Groupe Long Terme sur l'energie (Report of the Long Term Group on Energy), Paris, La Documentation Française, 1983).

Desprairies P, "Vingt ans d'industrie française du pétrole", Revue Française de l'Energie, Oct-Nov 1969, No. 215, pp. 124-133.

DGEMP (Ministère de l'Industrie).- Rapport du groupe de travail sur la réforme de l'organisation électrique et gazière française (rapport Mandil).- Paris, 1994.

DGEMP - Le chauffage électrique en France : étude historique, technique et économique, Paris, Ministry of Industry, 1988

DIGEC - Les coûts de production de référence : production électrique d'origine thermique (Reference Costs of Electricity Produced by Thermal Process), Paris, Ministry of Industry, 1993.

Energie Plus, no. 117, November 1992, p. 10-12

Finon, D L'échec des surgénérateurs: autopsie d'un grand programme. Grenoble, PUG, 1989.

Groupe Interministeriel sur l'effet de serre (GIES).-GIES Report.-Paris, Ministry of the Environment, November 1990, 93

Hayward, J The State and the Market Economy, Brighton, Wheatsheaf, 1986.

J.M.Chasseriaux, B.Chateau, B.Lapillonne - Un scénario de croissance sobre en l'énergie. -(A Scenario of Sober Growth in Energy), Report of the Ministry of Industry, Paris, La Documentation Française (Les dossiers de l'énergie, No. 21), 1979.

Lucas, N Energy in France: Planning, Politics and Policy, London, Europa, 1977.

MacArthur, J and Scott, B L'industrie française face aux plans (French Industry and the French Economic Plans), Paris, Ed. des Organisations, 1970.

Martin, J M, Criqui, P and Finon, D "La politique énergétique de la France depuis la première crise pétrolière" (Energy Policy in France since the First Oil Crisis). - in : Economica delle Fonti di Energia, No. 22, 1984.

Martin, J M, "Le secteur de l'énergie en France : un demi-siècle de profondes transformations" (The Energy Sector in France: A Half-Century of Profound Transformations). Zeitschrift für Energie wirtschaft, 14 (3), Sept. 1990, p. 224-237.

Martin, J M "Etat et entreprises énergétiques" (The State and the Energy Enterprises), in L. Nizard, ed., Planification et Société (Planning and Society) - Grenoble, PUG, 1974, p.121-140.

P. Brana, - Maîtriser l'énergie : un enjeu des années 90, - Report to the Prime Minister, Paris, , La Documentation Française, June 1989, 158 p.

Pean, P and Serenti, J P Les émirs de la république, Seuil, 1982

Radanne, P and Puiseux, L L'énergie dans l'économie, Paris, Syros, 1989.

Revue de l'Energie, n° 429, April 1991, p. 256-258.

Saumon, A and Puiseux, L "Actors and decisions in the French Energy System", in L. Lindberg, ed. The Energy Syndrome, Lexington, Lexington Books, 1977.

Shonfield, A Modern Capitalism: the Changing Balance of Public and Private Power, Oxford, Oxford University Press, 1976.

Simmonot, P Les Nucléocrates, Grenoble, PUG, 1978.

Souviron, J P Débat national Energie et Environnement, Rapport de Synthèse.- Paris, La Documentation Française, December 1994.

Thomas, S, The Realities of Nuclear Power: International Economic and Regulatory Experience, Cambridge University Press, 1988.

Toromanoff, M Le drame des Houilleres, Paris, Seuil 1969

3. German Energy Policy in Transition

Eberhard Jochem, Edelgard Gruber and Wilhelm Mannsbart
Fraunhofer Institut für Systemtechnik und Innovationsforschung (FhG-ISI)
Karlsruhe
Germany

3.1 Introduction

Unlike most other members of the European Union, Germany is a federal state. Accordingly, energy is not just a matter for the central government; the Land governments and even local councils are closely involved in energy matters. At the federal level, energy policy is considered a part of general economic policy, and is dealt with by the Federal Ministry of Economics (though in some Länder it is the responsibility of other ministries such as Environment or Public Health). The major objectives of federal energy policy are "to ensure the economic, efficient, secure and environmentally acceptable supply and use of energy". Federal energy policy, moreover is based on the assumption that the free market, subject to regulation on environmental and anti trust grounds, is the best way to attain these goals (BMWi 1990; IEA 1990). Indeed, with the exception of the protection given to the domestic hard coal industry (which is granted for reasons of social policy and supply security), the free market orientation of German energy policy has led to market-based pricing for most energy forms in Germany.

These goals and rules may be correct in principle, but do not take account of the high external costs of energy conversion and use - of the order of 30 to 50 b. DM per year as a result of corrosion, forest disease, agricultural losses, damage to human health and long-term climate changes (Prognos 1993). Indeed, there is a major contradiction in national policies stemming from the divergent approaches and concerns of the Ministry of Economics, the Ministry of Transportation and the Ministry of the Environment as well as of several Länder Governments. Contradictions are also visible in the contrast between, on the one hand, the Federal Government's ambitious target, set in 1990, to reduce West Germany`s CO_2-emissions by 25 % between 1987 and 2005 and, on the other, the rather modest energy and transportation policies implemented by the Federal and Länder Governments since 1990.

Germany is relatively poor in energy resources, importing about 98 % of the oil consumed, 78 % of the natural gas used and 100 % of uranium used. Total energy imports supplied two thirds of total primary energy demand in 1993, an increase proportion compared with the past (1980: 53 %). However, the real future constraint upon energy use seems to be the limited global absorption capacity for energy-related greenhouse gas emissions. If nuclear energy is not considered politically acceptable in Germany because of its extremely high damage potential and if the use of renewables is still relatively expensive, an "energy efficiency revolution" with least cost seems to be the answer for the next two decades.

The German economy's position, especially its links with the European Union and foreign markets weakens and slows down the government's scope for practical action: if a German environmentally-oriented energy policy develops too much ahead of the policies of the other EU member states (and even of other OECD member states), German industry will become less competitive in international markets, according to the associations of energy-intensive industries in Germany. The future single European energy market is beginning to have effects in a number of areas (e.g. energy conservation policies by the European Commission and competition in the utility sectors). Such policies may adversely affect the development of energy-related environmental policies, thereby imposing high social costs, at least from the West German point of view (Brand/Jochem 1990).

3.2   Energy Trends in West Germany and the Former GDR

Primary Energy Trends

Between 1980 and 1994 West German GDP (1991 prices) grew by about 34 %. Primary energy requirements, however, grew only by 5 %. The oil share dropped from about 48 % to 40 % and dependence on OPEC oil has also fallen for the present, though it is expected to increase in the medium term. Reliance upon natural gas and nuclear energy have brought about a diversification of energy supplies over the 1980s. The long-term decline of the share of domestic hard coal was interrupted in West Germany after the second oil price increase, when the mining companies and the federal government, expecting further oil price increases, revived the sector. The support given to the coal industry was reinforced by social policy considerations in the coal mining regions, the Ruhr and the Saar, which suffered from increasing unemployment rates already in the late 1970s.[1] The "revitalization" delayed the decline of coal's share in energy balances for almost a decade. In 1993, about 2.5 % of West Germany's primary energy consumption was supplied by renewable energy resources, with hydropower making up about 60 % of this total.

Table 3.1: **Primary Energy Consumption in Germany, 1980-94.**

Mtoe	1980		1987		1994 (prel.)	% Change 1980/94[e]
	West	East	West	East	Germany	
Primary energy requirements	273.1	85.2	271.6	93.7	335.8	-6.3
(percentage shares)						
- oil	47.6	17.3	42.1	13.0	40.4	-6.1
- hard coal	19.8	6.7	19.5	5.2	15.5	-12.8
- lignite	10.0	62.8	8.0	67.6	13.2	-45.3
- natural gas	16.5	8.5	16.8	9.5	18.3	17.4
- nuclear	3.7	4.0	10.8	3.3	10.0	25.0
- hydro[b]	1.9	0.6	1.8	1.3		
- other[c]	0.5	0.1	0.9	0.1	2.6[d]	19.8
GDP in 1991b.DM	2, 018	n.a	2, 218	n.a	2, 966	34.2[f]

a: not corrected for influences of weather and consumer stock variations

b: including net foreign electricity trade c: wood, wastes, landfill gas, etc.

d: for 1994 hydro and other are combined e: East and West Germany

f: only West Germany.

(Sources: BMWi; 1994; Statistisches Bundesamt)

---

[1] Employment in the West German hard coal mining industry stabilized at about 185 000 between 1978 and 1982 before it started dropping again to about 100 000 in December 1994.

On average, the intensity of primary energy requirements decreased by 1.8 %/year between 1980 and 1994. In West Germany, of course, these changes are not entirely due to improvements in energy efficiency; the influence of technical changes in energy-intensive industries and of long-term structural changes between the energy-intensive and less energy-intensive sectors of the West German economy also contributed to the decline in energy intensity. In the private consumer sector the trend towards more energy-intensive consumption, however, was strengthened between 1983 and 1987 when real income increased continuously (as reflected in increasing use of central heating, more single family houses, bigger dwellings and bigger cars). Generally, the decline in oil and other energy prices in 1986/87 seems to have slowed down the rate of efficiency improvements in several energy sectors (IEA 1990) because of extended pay-back periods for fuel conservation investments and because the potential for quick, highly profitable energy-saving investments seem to have been exhausted in the 1980s.

East German primary energy requirements decreased by about 41 % between 1980 and 1994, largely due to a sharp reduction in the population since reunification in 1990 with a loss of 800.000 inhabitants (-5 %) and a drastic economic breakdown in the East German industry (-50 %). In addition, the fuel structure changed drastically with declining use of lignite and hard coal and the phasing out of nuclear energy at the beginning of the 90s. On the other hand, the gas and oil requirement increased by 18% and 22% respectively. In West and East Germany together, primary energy requirements decreased by about 6 % between 1980 and 1994.

Final energy consumption fell by 6 % between 1980 and 1994 in Germany (see Table 2). While two sectors (commercial/public and industry) declined between 8 and 32 %, the transportation sector increased its energy consumption by more than one third. Industrial final energy use almost stagnated since 1984 in West Germany and declined by two thirds since 1990 in East Germany. In the case of transport, the following factors promoted this upward trend in West Germany: a sharp decline in fuel prices since 1986, an increase in the total number of bigger and more powerful cars and increasing mileage of cars, stagnation in the previous trend of improved new car fuel efficiency and the discontinuation of the trend towards diesel-driven passenger cars. In East Germany vehicle use sharply increased between 1990 and 1994 (up 40%).

Table 3.2: **Final Energy Consumption in Germany, 1980-93 (preliminary)**

Mtoe	1980		1987		1993		%Change '80/93[a]
	West	East	West	East	West	East	
Final energy consumption	179.8	54.1	179.7	57.7	187.4	32.3	-6.0
- Households	48.1	11.4	51.6	13.5	51.9	8.4	1.2
- Services	30.2	12.9	31.0	622	31.1	8.6	-7.7
- Transport	39.9	5.4	44.6	232	54.0	7.9	37.2
- Industry	61.6	24.5	52.5	998	50.4	7.4	-32.8
Electricity use (all sectors)	26.6	6.3	30.3	7.5	32.8	4.3	12.6
District heating	3.9	3.8	4.8	5.3	5.1	3.8	16.5

a: East and West Germany

Source: BMWi; Arbeitsgemeinschaft Energiebilanzen

Electricity consumption grew substantially in West Germany, but growth rates have slowed down in recent years and electricity intensity in 1994 was 8 % below the 1980 value. The installed capacity (1993: 103 GW) is sufficient to meet total electricity requirements (peak load in 1993: 58 GW). With the completion of some major power plants currently under construction, there should be sufficient capacity to meet demand in the coming decade. In East Germany electricity consumption (all sectors) decreased by 43 % between 1987 and 1994, thereby permitting the phasing-out of nuclear power stations which were considered to be not safe enough under the West German regulations.

In 1993, district heating accounted for about 2.7 % of total final energy consumption in West Germany but for about 12% in East Germany. Most systems in West

Germany are powered by coal and natural gas cogeneration plants whereas the East German district heat often was produced by lignite-fired boilers. Small engine-driven cogeneration plants have experienced a high diffusion rate during the last few years with a total of more than 1150 plants and total capacity of about 600 MW in use today (VDEW 1993). Small and medium-sized firms, contractors and municipalities are investing in this technology. In the renewables field, wind power has experienced a substantial growth reaching 650 MW by the end of 1994.

3.3 Energy Efficiency Policy in West Germany Over the Last Twenty Years

There are many similarities in energy efficiency policy between the Federal Republic of Germany and other industrialised countries: energy conservation programmes were started for most final energy sectors after the first oil price shock in the mid-seventies and were modified and strengthened after the second oil price crisis in the early eighties. Energy efficiency in West Germany has always been understood by the federal government as a market-oriented policy. Above all, it was assumed that world energy prices would give sufficient signals to the consumers. However, this was only true up until the beginning of the eighties.

Since the mid-1980s, decreasing fossil fuel prices have "coincided" with a weaker energy efficiency policy at the federal level (Gruber, 1992; BMU, 1994; see Figure 1). All financial incentive programmes for energy efficiency (except one low-interest loan programme for small and medium-sized companies) were discontinued between 1983 and 1991, first of all the 4.35 billion D-Mark programme for energy saving investments in private households (1978 to 1982) and a major grant scheme for district heating and cogeneration (1975 to 1977 and 1982 to 1986). Investment grants for industry as well as the tax reduction scheme for private households were phased out in 1989 and in 1991 respectively. Total subsidies for energy efficiency programmes in West Germany amounted to 13 billion D-Mark between 1973 and 1988 (Enquête-Commission, 1990c). According to the assessments made by Ifo (1990), all these programmes together resulted in primary energy savings of about 31 mtoe per year; including the effects of increased energy prices, the savings were estimated at 43 mtoe per year (i.e., some 15 % of total primary energy demand).

A joint Federal/Länder subsidy programme of 1.2 billion DM between 1992 and 1995 calls for repair and modernisation of district heating grids in the new Federal Länder with a particular emphasis on cogeneration.

In the case of consumer goods, an effective energy conservation policy has been carried out on the basis of voluntary agreements between government and manufacturers, as in the case of fuel efficiency of cars and electricity-saving of electrical appliances. These agreements were negotiated in 1978; the targets set for 1985 were even reached ahead of schedule. New agreements in 1985, however, were less goal-oriented, not mentioning any quantitative target over a given period of time. But four weeks before the first follow up conference on the Framework Convention on Climate Changes in Berlin in April 1995, the same two associations approached the government to offer negotiations about new voluntary agreements on fuel efficiency of cars and electrical appliances with quantitative targets. At the same time some major energy-intensive branches of the German industry offered negotiations on voluntary agreements with a target of a 25 % reduction of specific CO_2-emissions until 2005 (BDI, 1995).

Regulations exist to reduce energy consumption for space heating purposes: there are building codes for new buildings and standards for heating equipment. These were set up in 1977/78 and enforced in 1984. In comparison with northern European countries, however, building codes in Germany are still not very strict. Stricter building codes were discussed in 1992, and approved by the cabinet in mid 1993 and came into force in 1995 (the efficiency improvement is around 25 % for new buildings). Plans call for an additional 25 % increase in thermal insulation requirements in the late 1990s (Federal Environment Ministry, 1993). The introduction of individual heat metering in multi-family houses in the early 1980s has proved very effective: on average 15 % of space heat was saved. Concerning the electricity supply industry, a new federal tariff regulation came into force in 1990 which promotes the increase of the energy charge and the decrease of the fixed share of the electricity price thus increasing the profitability of electricity saving investments and encouraging energy saving behaviour. A similar tariff regulation for gas and district heat has been proposed (Brand et al., 1988), but the issue has not been taken up by the administration.

Although the Federal Government states that information, education and consultancy "have special significance", the advisory scheme for private consumers was stopped in 1993, the information campaigns operate at a low and sporadic level and the initial consulting programme is only used by a few hundred small and medium-sized companies. The new subsidy scheme for energy audits in buildings has not been evaluated so far.

Fuel and oil taxes have been in place in West Germany for many years, even before the first oil price shock, and are therefore regarded by energy consumers as a part of the energy price. In 1989, the oil and fuel tax was slightly increased and a small gas tax was introduced. The price effects of these measures, however, were within the range of usual fluctuations of the market prices. In July 1991, these taxes were again slightly increased, though for the sake of the federal budget rather than for energy conservation. A further tax increase on gasoline and diesel become effective at the beginning of 1994. A tax benefit for oil burning cogeneration plants has been decided within the Mineral Oil Tax Law in 1992.

In the transport sector, there are hardly any programmes for energy conservation or reversing the trend of the modal split. The increase in total fuel consumption in this sector by 70% since 1973, despite the improvements in fuel efficiency, is obviously a consequence of this failure.

The reduced energy efficiency policy at the federal level (which coincides with the decline of the oil price in 1986) was reflected in reduced energy conservation efforts in all sectors, with the exception of energy-intensive industries and new buildings.

Renewable Energies

Most efforts using financial incentives were concentrated in the field of renewable energy (e.g. the demonstration programmes "250 MW wind" and "1000 roofs photovoltaic") in 1991-1994. The subsidy for wind power is 8 DPf/kWh produced and runs for 10 years. In addition, the buy-back rates for electricity from renewables had been fixed as a percentage (70 % to 90 %) of the electricity rates of end users by the Act on Sale of Electricity to the grid, taking effect in January 1991. Environmental legislation on gas emissions from landfills and on sewage treatment plants induced engine driven cogeneration using landfill and digester gases.

Länder Governments and Local Authorities

The Federal Ministry of Economics' modest commitment to energy efficiency policy - the key policy to meet the target of 25 % CO_2 emission reduction by the year 2005 - has partly been compensated by several factors: an active waste management legislation, initiated by the Federal Ministry of Environment, structural changes of energy-intensive material production (steel, aluminium, chemicals), policy initiatives at the Länder Government level and by local authorities as well as first changes of the philosophy of utilities to develop energy services.

Recognising the existing imperfections of energy markets, the numerous obstacles to energy efficiency, the social cost of energy use as well as their own responsibilities in this field, some Länder governments and many local authorities have undertaken substantial initiatives. Many of the 16 Länder governments have started energy efficiency programmes for selected target-groups (e.g. information, motivation, initial consulting, and contracting). They are pursuing this policy in a very active manner, notably in North Rhine Westphalia, Saarland, and more recently also Schleswig-Holstein, Lower Saxony, Hessen and Brandenburg. For example, they have introduced energy agencies which apply a group-specific approach in order to alleviate existing obstacles in small and medium-sized companies or communities. These agencies are trying to act as "catalysts" and "brokers" for energy efficiency investments (e.g. initial consulting, professional training, contracting). Some of the energy agencies cooperate with utilities or with Länder banks. Other activities of the Bundesländer are education and vocational training in the field of energy efficiency, specific consumer information, energy efficiency diagnosis of buildings, and price policy on electricity tariffs (Clausnitzer/Hille, 1993).

An increasing number of local authorities are also developing their own energy efficiency and transportation policy programmes which aim at reducing greenhouse gas emissions by 25 % or more (the "Climate Alliance" cities). The authorities are trying to attain their objectives by designing and implementing local and regional energy demand and supply concepts which concentrate on waste heat recovery, the promotion of combined heat and power generation, the utilisation of digester gas from sewage treatment plants and gas from landfills, local transport planning concepts and possibilities of reducing space heat demand. They also try to improve information and advisory services offered to citizens. The policy of separating municipal waste at source in private households and collecting three or four separate fractions (e.g.: glass/ wood, plastics, metals, paper/ biodegradables/ others) increases the potential for improved recycling of energy-intensive materials and hence, energy efficiency in industrial production (Federal Environment Ministry, 1995)

In addition, an increasing number of municipalities (mainly those with "Stadtwerke", such as those of Saarbrücken, Hannover, Kassel, Munich, and Rottweil) have been implementing measures of energy services and the first tentative elements of least cost planning. These activities were often initiated by city councils where the Social Democrats or members of the Green Party were in the majority. This policy also seems to have had an influence on the nine major electricity utilities. RWE, the biggest utility in Germany, established a 100 million DM bonus scheme for consumers purchasing energy-efficient household appliances.

Recent studies (Gruber and Brand, 1991; Jochem and Schäfer, 1992) show that there is still considerable potentials for profitable energy efficiency investments despite the current relatively low energy prices. German energy policy needs to apply a bundle of measures to overcome the existing barriers (Jochem and Gruber, 1990). Activities such as motivation of energy consumers, target group- oriented information and training, consultation, new financing models (such as contracting) to overcome financial restrictions and the investor-user dilemma, the use of trustworthy intermediate institutions such as trade associations, as well as price signals set by the government and the provision of a suitable infrastructure and legislative boundary conditions all have to be coordinated and carried out simultaneously in order to realise existing profitable efficiency potentials.

3.4 Energy Supply Policies and the Impact of Environmental Legislation

The German energy supply sector is organised on a private enterprise basis.[2] While the West German Federal government has in the past held stakes in industrial holding companies with energy subsidiaries (such as VEBA and VIAG), these were sold off in the 1980s. In some cases, Länder or municipal governments hold shares in electricity and gas utilities, although these too are in decline (Schiffer, 1988).

---

[2] The East German energy sector was of course publicly owned.

Nuclear Energy

In the late 1960s and early 1970s nuclear power was considered by all political parties as the most promising future primary energy source in West Germany (currently, some 30 % of German electricity consumption is generated by nuclear power stations). In the 1980s the energy R&D budget was dominated by nuclear power (although its share was declining). However, over time opposition to nuclear power has grown, mainly on the basis of fears over the potential damage which would be caused by a major accident in highly populated areas (regardless of its low probability), the necessity for safe storage of nuclear wastes over many centuries and the potential military misuse in politically unstable countries.[3] In 1986, shortly after the Chernobyl accident, the Social Democratic Party decided to phase out nuclear energy use within 10 years. Although a decision was not introduced in the German Parliament, utilities and manufacturers associated with nuclear power face an uncertain future. In 1989 the nuclear industry decided to discontinue construction of the fuel reprocessing plant in Wackersdorf. In 1990 the government of North Rhine-Westphalia stopped operation of the high temperature reactor, a 300 MW demonstration plant and, in 1991, the fast breeder demonstration reactor which was ready to be activated. In the same year, the four operating nuclear power stations of the former GDR were taken out of operation for safety reasons.[4]

Although there is no formal policy decision on nuclear power use, a moratorium for the 1990s exists in practical terms which, of course, excludes the utilities from using this technical option to reduce their high greenhouse gas emissions. Meanwhile, the utilities and the nuclear industries of Germany and France are developing a new generation of nuclear power stations, in which the potential maximum release of the nuclear materials and the probability of a major accident will be reduced by several orders of magnitude. They plan to put these new stations into operation when the technical lifespan of the existing stock is terminated (Unger, 1992), which will be after the year 2010. This plan is in conflict with the current attitude of the Social

---

[3] Potential damage of a major accident has been estimated in the range of 4 to 11 trillion DM, two to four times the German gross domestic product (Ewers, 1991; Moths, 1993).

[4] Although the West and East German grids were not interconnected in this period, the shutdown did not cause any shortages because East German industrial demand for electricity fell dramatically due to severe structural changes and production cuts.

Democrats and the Greens towards nuclear energy, which they now want to be phased out during the technical life span of the existing stock.

In the autumn of 1992 a few of the big utility companies declared their unwillingness to invest in the further development of a new generation of nuclear power plants as long as a consensus on the future role of nuclear energy does not exist between the Christian and Social Democrats. The Christian Democrats and the German Industrial Associations (BDI and DIHT) are in favour of nuclear power because of their belief in the technical progress of nuclear safety, the manageability of safe long-term storage and non-proliferation and the competitive electricity production costs. The German Industrial Association has been complaining about the high electricity prices for industry relative to those in other countries such as France, the Netherlands or the United Kingdom for many years (BDI, 1992), which it regards as a substantial competitive disadvantage for German industry.

A consensus on future energy use (including nuclear energy) was the objective of a Commission of politicians and other representatives formed in the first half of 1993. Although a consensus which covered performance criteria for future nuclear power plants, commitments for an engaged policy on energy efficiency and reduction target for domestic coal production seemed to be possible, the Social Democrats stopped the negotiations. But all parties declared their will to continue the negotiations which are planned to start again in summer 1995.

Hard Coal

The West German hard coal industry has experienced a steady decline in production since 1956 due to unfavourable and very costly deep mining conditions. In the course of this adjustment, the number of working pits has shrunk from 175 in the mid-fifties to 28 at present. The total number of workers in the coal sector decreased by more than 80 % from 600,000 to 100,000 in December 1994. Considerable strains and unemployment in the mining regions followed, particularly in the Saar and Ruhr areas. This was addressed with extensive public subsidy, mainly for social and political reasons, amounting to 10.7 billion DM in 1990 (Heilemann and Hillebrand, 1992). The intervention per tonne, 60 ECU for West German hard coal in 1992, is high compared to figures for France and Spain (20 to 26 ECU per tonne).

The use of domestic hard coal is concentrated in electricity production (41 Mill t per year) and in the iron and steel industries (about 20 Mill t per year). The remaining 7 Mill t (in 1991), which is used by other branches of industry and exported to member countries of the EC, will shrink rapidly to negligible quantities in the 1990s. Despite substantial progress in labour productivity (between 1982 and 1992: 2.8% per year), domestic hard coal is three and a half times as expensive as imported coal. To protect the domestic coal sector from cheaper imports a rather complex system of laws and private contracts has been established during the last 20 years, which aims to secure sales to the electricity and steel industry.

The use of coal in the power industry is regulated by a formal contract in which the utilities have committed themselves to buy a fixed amount of domestic hard coal of 40.9 m.tonnes per year until 1995 and 35 m.tonnes per year until 2005. These private contracts are backed up by governmental subsidies, mainly financed by a special levy, the so-called "Coal Penny", equivalent to 7-8% of electricity prices. The contract was negotiated and agreed upon by the two industries and the Federal Government largely on social policy grounds (though it was also justified as supporting supply seucrity). But there were political negotiations and pressure from German industry to put the financial burden on the tax payer by 1996 (BDI, 1992). In December 1994 the Federal Constitutional Court declared the levy unconstitutional and required its abandonment by the end of 1995.

Public intervention in the German iron and steel industry took the form of delivery and purchase contracts between the coal and steel industries, complemented by guidelines and obligations of the federal government, the mining Länder and the EC Commission. These subsidies totalled 3.7 thousand million DM in 1990.

A commission of the Government of North Rhine-Westphalia on future coal policy (1990) recommended a reduction of West Germany's hard coal production from today's 65 Mill. tonnes per year to 40 or 30 Mill. tonnes by 2010. In Germany the so-called "Kohlerunde" ("Coal Round Table") agreed in 1991 to reduce domestic coal use for the electricity and steel industry from 65 million tonnes in 1991 to 50 million tonnes in 2000. This reduction would allow a cutback of subsidies to 7.5 thousand million DM per year and would lead to the layoff of around 30,000 miners or 3,000 per year which seems politically acceptable. Reduced iron and steel production, improved energy efficiency in iron production and electricity use and a partial exodus of electricity-intensive production of aluminium, ferro-alloys and chlorine may accelerate the decline of domestic coal production in the 1990s to some 40 to 45 million tonnes in 2000.

An interesting effect of the German coal policy is the fact that it stabilises the nuclear energy capacity: the production of hard coal electricity would be cheaper than the production of nuclear power not only for medium load but also for base load, if the German utilities had free access to international coal markets (Heilemann and Hillebrand, 1992). It could be argued that a phasing-out of the German subsidies for domestic coal would have adverse consequences not only by increasing unemployment by 150-200,000 but also by making the reduction target of CO_2 emissions by 2005 even more unlikely due to increased power production based on coal imports and lower electricity prices. This latter effect demonstrates the risks of pursuing a free market policy as long as external costs are not internalised.

Lignite

Lignite, which is very CO_2-intensive, is the major domestic and internationally competitive energy source in Germany (especially in East Germany). Almost all West German lignite is used for electricity generation in heat power plants with efficiencies below 40 %. Lignite is extracted from huge open mining fields (120 m.tonnes per year) entailing the destruction (and eventual restoration) of landscapes in densely populated regions and a substantial decrease of ground water levels. Although these external costs are increasing, it is doubtful if production will be reduced in the next two decades. In Rhineland, the preparation of a new mining field (Garzweiler II) has proved highly controversial (though it looks likely that it will go ahead despite political changes in the Land).

Oil and Natural Gas

Oil is still Germany's main fuel supply (40 %, see Table 3.1). In this energy sector largely free of government intervention, about 50 companies are involved of which about 15 are companies with their own refining capacity. Some are subsidiaries of international, European and American oil companies while a few are subsidiaries of German companies (e.g. VEBA Oil, DEA). Since the mid 1970s oil refining capacity has declined sharply due to reduced demand (see Table 3.1) and structural changes in European and Near East refinery capacities.

Some oil products are delivered direct to the consumer by the refinery companies, but heating oil is delivered mainly via the wholesale trade and many independent retail companies. Whereas total final consumption of oil products declined by 4 % between 1970 and 1990 the structural changes among the oil products was substantial: vehicle fuels (gasoline/diesel etc.) increased by 83 % while heating oil fell by 45 %. Heating oil use was reduced due to efficiency improvements and increased use of natural gas and district heat from coal or natural gas.[5]

Natural gas became the second largest primary energy source in Germany in 1992. Growth is still very high in the residential sectors, where 80% of new built dwellings are heated by natural gas. There has not been much government intervention in this energy sector except a 30 % limit on gas imports from the former Soviet Union (which was removed in 1993 due to the political changes) and limits on gas use in power generation. Natural gas is promoted by many municipalities ("Stadtwerke") and with increasing interest by electricity companies. The natural gas business includes producers of domestic natural gas and gas importing companies (Ruhrgas AG with the highest market share), as well as regional distributors and municipalities. There is no direct competition between gas suppliers because regional and local distributors have long term contracts with their suppliers and with the local or regional authorities to supply within a defined area ("demarcation contracts") without any competition. Natural gas as a fuel competes - it is said - in the space heating sector with heating oil, electricity, district heat and coal. In practice, however, once the investment decision has been made, the influence of competition among fuels is weak.

Electricity

In Germany, electricity supply is generally structured into three sectors, very different in size and market influence: public supply; industrial generation; railway production.

---

[5] The possibility to compensate for the lost heating oil sales by energy services is still rarely taken up by the trading companies.

Public electricity supply covers sales to industry, private households, commercial/public, transport and agriculture across a public electricity grid and does not relate to the question of ownership. The "public" aspect is not an indication of state jurisdiction; on the contrary, the organisation falls under private law. There are three groups of electric utilities (Schiffer, 1993). The "Verbundunternehmen" (the "Big 8" in West Germany [6] and VEAG in East Germany) plan, operate and coordinate major power stations and a high voltage transmission network on a supra-regional level. The regional distributors (about 60 companies in West Germany with partially owned power stations) distribute to local companies or directly to final consumers. They are members of an organisation known as the ARE ("Arbeitsgemeinschaft regionaler Energieversorgungsunternehmen"). About 800 local municipalities ("Stadtwerke") in West Germany (not including municipalities founded in the territories of the former East Germany since unification) distribute directly to final consumers in their district (often together with gas, district heating, water and in some cases local public transportation); they form the association VKU ("Verband kommunaler Unternehmen") and in some cases have some generation capacity, mostly steam based or engine driven cogeneration. Most of the electricity generating and distributing companies are represented in the association VDEW (Vereinigung Deutscher Elektrizitätswerke).

Industrial electricity generation is focused upon around 200 companies in West Germany, organised into the association VIK, "Vereinigung Industrieller Kraftwirtschaft". Mostly based on cogeneration; (small hydro power plants play a very minor role), industrial supply contributes 13 % of total electricity consumption of Industry in West Germany. Electricity generation of the Federal Railways (contribution to total electricity use is only 2 %). The voltage and the frequency of the power for federal railways differ from the public grid.

Electricity supply in Germany is relatively decentralised, with companies of mixed ownership dominating electricity production (though there is a high degree of capital investment by the public sector at the level of the municipalities. However, while the number of utilities suggests that the industry has a rather pluralistic structure, there is a very high degree of concentration in electricity generation and high voltage transmission. This structure is not without tensions and conflicts of interest: the big

---

[6] It is expected that two companies in Baden-Württemberg, the Badenwerk and Energieversorgung Schwaben (EVS), will merge within the next two or three years.

eight in West Germany, for instance, sell electricity at prices based on average production costs (including sunk costs). This practice does not give regional distributors or municipalities much opportunity to build up their own generation capacities (on a marginal cost basis), except for cogeneration plants under favourable conditions.

The Energy Management Act (Energiewirtschaftsgesetz, EnWG), passed in 1935 and still valid today as a federal law is intended to provide a basis for safe and economical energy supply. The formulation of this law assumes that for the two grid-based energies, electricity and gas, optimal supply organisation is possible only within closed supply areas, and not from competition for customers. In the case of subsidiary distribution of electricity, the supply areas are delimited by "demarcation contracts" (Demarkationsverträge). Additionally, there are end user protected area contracts (Konzessionsverträge) between the utilities and the regional and local authorities which give the utility exclusive right of distribution to final consumers. The supply monopolies are justified by the special characteristics of electricity production and distribution (high investment intensity, long pay back periods, management grid dependency, impossibility of adequate storage; obligation to connect and supply).

In order to ensure state influence, the Energy Management Act provides far-reaching possibilities for influencing investments, market access, prices and business conditions. Permission from the public authorities is needed in order to start supplying third parties with electricity or gas. The utilities have an obligation to inform the relevant supervisory authority whenever they plan to build, extend or shut down power plants or transportation lines. The supervisory authority has the power to prevent the projects of which it has been informed in the interest of the common good. Thus changes in capacity are only possible following examination by the public authorities, which have a difficult task as they have only a few officials per Land Government. In theory, companies can have their operating permission taken away, though for the most part the regulatory system is too weak to make this threat credible.

The electricity prices for tariff users (small energy consumers) are subject to supervision by the public authorities (Bundestarifverordnung Elektrizität, Federal Regulation on Electricity Rates, revised in 1990). There is also a general federal regulation on gas rates. Tariffs can only be raised with the approval of the responsible price supervisory authority (the Ministry for Economics in each "Land"). The prices for bigger energy consumers are exempt from this supervision, but are subject to supervision according to cartel law (the 1957 Wettbewerbsrecht). This legislation has a wider significance for the network energy industries. In 1980 an amendment established that protected area contracts must be limited to 20 years. In 1990 cartel laws were modified to strengthen competition - for example, as with natural gas, by improving opportunities for users to buy electricity from other producers, under certain conditions, and to arrange "carry through" agreements with local authorities (IEA, 1991). In 1990 the basic ban on negotiating new licence fees and increasing licence fees - previously grounded in the "Konzessionsabgaben-Anordnung" - was found to be unconstitutional. Since then, communities can no longer be prevented from negotiating new licence fees or raising already existing licence fees.

3.5 Environmental Policy in the Energy Sectors

Conventional power stations, gas works, combined heat and power stations and boilers are subject to regulation and approval under the Federal Pollution Control Act (Bundesimmissionsschutzgesetz). In some German Länder there are also regional pollution control laws, specifying e.g. limitations upon certain types of fuel in conditions of high atmospheric pollution.

The government's environmental policy was rather successful in the various energy sectors in the 1980s. Regulations on desulphurisation and removal of NOx from stock gases were imposed on boilers by the ordinance on large boilers (>100 MW) and the technical ordinance on air emissions ("TA-luft" covers boilers from 1 MW to 100 MW). Due to these regulations, and to a lesser extent to fuel substitution in industry, West German SO2 emissions from the energy sector and industry were reduced from 2.8 m.tonnes in 1980 to 0.75 m.tonnes in 1989. NOx emissions from these sources underwent a 40 % reduction in the same period and will be reduced still further in the 1990s by this legislation (UBA, 1992). Financial incentives for catalytic converters in existing cars and regulations on new cars contributed to a substantial decrease (31%) of CO and a stagnation of NOx and hydrocarbons emissions in the 1980s (even though the fuel consumption of cars and trucks was still increasing); in East Germany lower emissions were mostly due to a lower proportion of cars per 1000 inhabitants (in comparison with West Germany). Environmental legislation also aimed to reduce specific emissions of CO, CO2, hydrocarbons and NOx for small boilers and burners by an ordinance on small firing installations which sets efficiency standards. It also introduced an emission label (the so-called "blue angel") which guarantees the buyer of new burners emissions below a maximum standard level negotiated between the manufacturers and the Environmental Protection Agency in Berlin.

3.6   Energy Forecasts 2010 and 2020

According to the most recent energy forecasts for the German Government (Prognos/IfE, 1991), the primary energy requirements of West Germany are expected to increase by about 5 % between 1990 and 2010, a comparable change has been experienced in the West German population due to migration from East Germany, Eastern Europe, the Soviet Union and Asia.[7]   Total final energy consumption of West Germany is expected to increase by nearly 6 % between 1990 and 2010, with a 4 % increase in the residential sector (including a 10% increase in energy used in space heating), and a 15 % in the commercial/public sector (with electricity consumption growing by 42 % and other fuels by 7 %). The energy demand of the West German industrial sector is expected to increase by 8 % (electricity up 28 % compared with 1% growth for other fuels).

The transportation sector remains at about the 1987 level, on the basis of a higher share of diesel-driven private cars, reductions in the distance travelled by cars/year and further technological improvements. However, the trends in road transportation over the last few years seem to suggest that the transport demand projections may be the weakest point in the forecast if transportation policy does not change its present laissez-faire attitude. Between 1990 and 1993, energy demand in the West German transport sector rose by 8.2%, rendering it the largest energy-using sector.

The share of oil, hard coal and lignite in West Germany is expected to drop (see Table 3.3) because of further substitution of oil by gas for space and process heat and a reduction in coal use for electricity generation. Gross electricity consumption is expected to increase by about 18 % from 441 TWh (1989) to almost 550 TWh

---

[7]   About 4.4 million people (or 7 %) migrated between 1987 and 1994.

(2010). The share of total net electricity imports is expected to grow to about 5 % of gross electricity consumption. Renewable energy production is expected to grow from 2.5 to about 4 % of total primary energy requirements due to the modest energy price increases assumed and the various promotional programmes mentioned earlier, mostly in wind power, thermal solar and biogas.

Table 3.3: Energy Forecast for FR Germany (status quo projection), 1990-2010

	West Germany			Unified Germany		
	2010 (Mtoe)	2010 (%)	1990 (%)	2010 (Mtoe)	2010 (%)	1990 (%)
Total primary energy required	288.5	100.0	100.0	354.2	100.0	100.0
-Oil	107.0	37.1	49.0	128.9	36.4	38.4
-Hard coal	49.6	17.2	18.9	58.2	16.4	16.1
-Lignite	21.0	7.3	8.2	40.4	11.4	17.0
-Natural gas	64.6	22.4	17.7	79.2	22.4	16.8
-Nuclear	31.0	10.8	12.0	31.0	8.8	9.6
-Hydro/net trade	10.4	3.6	1.2	10.7	3.0	2.1
-Other renewables	5.0	1.7	1.0	5.7	1.6	
Total final energy consumption	187.2	100.0	100.0	231.3	100.0	100.0
-Households	46.1	24.6	25.0	56.6	24.5	25.2
-Services	33.6	17.9	16.5	42.8	18.5	18.1
-Transport	49.2	26.3	28.1	60.6	26.2	25.2
-Industry	58.3	31.1	30.3	71.3	30.8	31.5

Source: Prognos, 1991; Arbeitsgemeinschaft Energiebilanzen

The energy forecast for unified Germany shows a constant primary energy consumption between 1989/90 and 2010 (West: +5 %, East: -17 %) compared to a

74 % increase in GDP. Total energy intensity is expected to improve by 33 % (West), and 75 % (East). An increase in the share of oil products and of natural and other gas is expected in view of the structure of primary energy demand. On the other hand, the share of solid fossil fuels is expected to decline from about 39 % to around 30 % (2010) (see Table 3.3). The share of electricity in total final energy consumption is likely to increase to almost 20 % in 2010. As a result CO_2 emissions will decrease by about 8 % in Germany (5 % increase in western part and 34 % decrease in eastern part) due to massive reduction of lignite use in electricity generation, industry and the residential sector. The use of domestically produced lignite dominated the primary energy consumption in East Germany in the 1980s, e.g. the share of domestic lignite was 67 % in 1987.

The forecasts assume major improvements in the field of energy efficiency, including a tightening-up of thermal insulation standards for new buildings along with decreasing specific consumption of electric appliances in the residential sector, a 25% reduction in vehicles' specific fuel consumption and general reductions in consumption within industry (though for electricity consumption would rise in some sectors where new techniques or automation would require it).

In order to understand the forecasts made for the German Government, it should be noted that future energy policy decisions by the Government are assumed by the analysts and completely incorporated in the forecast: the German energy forecast is the most likely energy future seen by the analysts involved. This may mislead readers not familiar with the forecast's philosophy to suppose that the forecasted changes will come about without great effort or much alteration of policy, generally labelled as the „reference scenario". Another interesting point is that the authors of the Prognos/IfE forecast obviously did not expect the German Government to meet its own reduction target (-25 % CO_2 emissions in 2005) set in December 1990.

Two other more recent forecasts made by oil companies expect lower primary energy consumption and lower CO_2 emissions due to reduced economic growth assumption, intensive structural changes in the economy and greater use of natural gas. On the other hand, the projection made by the Commission of the EU in 1992 expects 13 % increase in primary energy demand and a 9 % increase of CO_2 emissions, but on a per capita basis a stagnation of per capita CO_2 emissions by the year 2005. This seems to be primarily a reference scenario, excluding specific energy policy action.

3.7 The New Climate-related Energy Policy in Unified Germany

The German Physical Society became concerned about the impacts of climate change in the early 1980s. At that time, a strong opposition to nuclear energy had developed in the political and scientific communities. High energy demand projections and a practical moratorium on nuclear energy created deep concern about future growth of fossil fuel usage. The International Conference in Villach 1985 came up with such serious and repeated warnings on climate change and its impact that, following an initiative from a few scientists and parliamentarians, the German Bundestag set up the Enquête Commission on Preventive Measures to Protect the Earth's Atmosphere in 1987. The Enqête Commission (which had to report to the Bundestag and to complete its mission by mid-1990) divided its task into three major areas: the depletion of stratospheric ozone, the situation of the tropical and boreal forests, and the role of energy-related greenhouse gas emissions. The Enquête Commission held numerous hearings and ordered more than 100 studies on energy issues, energy-related emissions and their reduction potentials, and international climate change policy measures (Enquête Kommission, 1990a). On the basis of this evidence, the Commission prepared a synoptic energy report (Enquête Kommission, 1991).

In order to determine the potential for emission reductions (and their costs) the report developed three scenarios for the period 1987-2005, based on different assumptions for the development of nuclear power, natural gas, and renewables. Besides the "trend savings" of the reference scenario on the order of 165 million tonnes CO_2, an additional 120 million tonnes of emissions could be avoided by more rational use of energy (see table 3.4). The order of magnitude of the contributions of the three substitution options - renewables, intrafossil substitution, and existing nuclear power plants at full load - by 2005 is about the same in the "energy policy" scenario: 25-30 million tonnes CO_2 in each case. Adding a small contribution from modifications in consumer behaviour, a 30 % decrease of CO_2 by 2005 seemed achievable.

On the basis of these findings, the Enquête Commission recommended a 30 % decrease in CO_2 emissions as a national reduction goal (referring only to the territory of West Germany before unification) by 2005 compared to 1987 levels. A reduction of global CO_2 emissions by at least 50 % by the year 2050, which the Enquête Commission considered to be necessary under current conditions and knowledge in order to limit the anthropogenic contribution to the greenhouse effect, would mean that the industrialised nations would have to reduce their CO_2 emissions by 80 % by the year 2050.

The Enquête Commission also adapted the reduction targets for unified Germany, requiring CO_2 emissions to be reduced from about 1070 million tonnes in 1987 to 750 million tonnes by the year 2005. The Enquête Commission knew that its recommended targets and measures for achieving those targets were rather radical. Minority votes on particular measures and technology options underline the difficult task within a 15-year period.

A similar recommendation was made in a study conducted jointly by the German Federal Ministry for the Environment, Nature Conservation and Nuclear Safety (BMU) and the German Federal Environmental Protection Agency in 1990. This study served as a basis for the decision made by the German Federal Cabinet on June 13, 1990 to the effect that energy-related CO_2 emissions should be reduced by 25 % by 2005 relative to 1987. An interministerial task force was commissioned to work out and submit proposals for measures designed to achieve this reduction target. The Federal Cabinet reaffirmed its target on November 7, 1990 for West Germany and declared a higher percentage reduction („25 %+") in the new federal states (i.e. the former GDR) in view of the substantial potentials of energy efficiency and lignite substitution attainable there (Fed. Environment Ministry, 1991).

The issue was found to be so important that the 12th German Bundestag decided to establish a second Enquête Commission on climate change during the period of 1991 to 1994. The Commission reconfirmed the CO_2-reduction target for 2005 in 1994, although it had to state that the energy policy initiatives of the Federal Government were far behind the time schedule the first Enquête Commission had envisaged as the targets were defined in 1989/1990 (Enquête Kommission, 1995).

Table 3.4: CO2 Reduction Potentials of the Enquête Commission's "Energy Policy" Scenario in 2005 in West Germany (based on 1987)

Major assumptions of the „Energy Policy" Scenario		
- no additional nuclear capacity, but increased operating hours		
- removal of obstacles to rational energy use		
- intensive policy to change modal splits in the transportation sector		
Composition of technical options	**M. tonnes CO_2**	**% share**
Rational use of energy		
- additional savings in final energy sector		
(including modal split policy)	-75.0	-10.5
- savings in the transformation sector and		
savings due to cogeneration	-45.0	-6.3
Renewables	-30.0	-4.2
Intra-fossil fuel substitution	-26.5	-3.7
Nuclear power plants at 85 % capacity	-25.0	-3.5
Energy-conscious behaviour	-13.0	-1.8
Total (compared to 1987)	**-215.0**	**-30.0**

Source: Enquête-Commission, 1990 b

Prospects for the Climate-related Energy and Transportation Policy

At first sight, these reduction targets seem to be too ambitious and many doubts about their feasibility have been expressed at public hearings and in comments from, among others, the energy suppliers and the Federal Ministry of Economics. On the other hand, Germany would only be reducing its per capita CO2 emissions rate from 13.7 tonnes in 1987 to 9.6 tonnes in 2005, which is, in fact, the current average within the European Union.

The total CO2 emissions of 1,065 m.tonnes of unified Germany in 1987 were reduced by 16 % to 890 m.tonnes in 1994 due to a number of reasons. The energy requirements of East Germany have drastically decreased since 1987, the base year for the Government's 25 % reduction target, from 93.7 mtoe to 49.8 mtoe, due to the economic break down and severe structural changes in the East German economy. East German lignite products were substantially substituted by oil products and natural gas while West German hard coal and oil products were partly substituted by natural gas. Practically all CO_2-reduction since 1987 stems from East German changes in energy use which cut CO_2 emissions in half between 1987 and 1994. This drastic decline - sometimes labelled as „wall fall profits" - hides the fact that CO_2 emissions of West Germany increased since 1987 by some 10m.tonnes (or 1.4 %). On a per capita basis, however, the change was minus 6.0 % because of an increasing population.

The per capita CO2 emission rate of 10.9 tonnes per year of unified Germany in 1994 is still very high in relation to today's average per capita emission rate of 9 tonnes per year for the EU, or the world average of 4.2 tonnes per year. The relatively high per capita CO2 emissions of Germany are due to a still substantial use of lignite and hard coal; their share in total primary energy requirements was 28.7 % in 1994 and this is expected to decrease further until 2010. If the German Government wants to meet its reduction target by 2005, the average emission reduction has to be 1 % per year in the future. Assuming an economic growth of around 2.0 % per year on average, the necessary reduction of CO2 emissions is about 3 % per year. One can argue that this tough target could still be achieved by extensive energy efficiency improvements and changes in the transportation policy, economic structural changes, interfuel substitution and less energy-intensive consumption patterns of private households. But many experts doubt in the mean time whether the Federal Government still really wants to meet the target.

3.8   The Impact of the European Union upon German Energy Policy

Despite these uncertainties, German concerns over the environment are arguably much stronger than in much of the rest of the European Union and it is no surprise that this has prompted the German government to press for European policies to take account of the environment. In part this is due to concerns over the competitiveness of German industry; if other countries are obliged to incorporate standards close to those agreed in Germany then not only does this limit the damage done to German competitiveness vis a vis its European trading partners but it also gives German industry a strategic advantage. Thus it is not surprising that the German government has been to the fore in seeking a Europeanisation of environmental standards. Arguably this was most successfully achieved in the moves to control emissions of gases contributing to the acid rain problem. Elsewhere, however, the policy may not have been so effective. In the negotiations on the reduction of the sulphur content in heating oil and diesel in 1990, the Commission did not accept the West German proposal to cut the sulphur content by 50 %.

However the German government has been less enthusiastic about a European dimension in other aspects of energy policy. The government is one of those most reluctant to see a European energy policy develop, preferring to retain autonomy in the pursuit of supply security. Indeed the supply security rationale has been to the fore in its support of the German coal industry and its hostility to the European Commission's attempts to monitor and control subsidies in this sector. Indeed, in contrast to its professed free market approach to the energy sector, the German government has been rather ambivalent about the liberalisation of energy utility markets both at home and in a European context. It has offered only half-hearted support to the Commission in its attempts to open up electricity markets while it was strongly opposed to similar moves in the gas sector. In both cases it appears that the German government was to some extent reflecting the interests (or conflicting interests) of the national industries.

3.9   Conclusions

Whatever the German Federal Government finally decides in order to promote more efficient energy use and to reduce CO_2 emissions by the remaining 90 m. tonnes within the next 10 years, it will have to reverse its present policy. The ratio of subsidies made by the Federal Government for energy supply (coal, gas, oil, nuclear) as opposed to improving energy efficiency varied between 6:1 and 9:1 during the 1980s. The termination of many grant schemes for energy conservation in private homes, for market introduction of energy-saving products, for district heat and coal-fired cogeneration as well as energy-saving investments in industry (see Figure 1) may be reversed in the next few years and the schemes reintroduced, although the financial burden of the unification of Germany may limit these possibilities.

When the German Federal Government defined its CO2 reduction target in 1990 it was aware of the particular difficulties involved in the transport sector. It therefore assumed that it would only be possible to stabilise transport-related CO2 emissions by the year 2005. The policy envisaged includes the following measures (Müller, 1992):

- The current motor vehicle tax is to be replaced by an emission-related tax with a CO2 component.

- CO2 emissions are to be limited for each vehicle class, the aim being to reach an average fuel consumption of 5 to 6 litres per 100 km for new cars (today's car stock consumption is 9 litres per 100 km). This measure can only be implemented within the framework of the European Community.

- The rail infrastructure is to be improved, particularly in the new Länder, by intensive investments, as is the attractiveness of local public passenger transport.

- A further increase in the gasoline tax and the introduction of a toll for cars using Germany's motorways are under discussion as are further speed limits on motorways (general speed limit) and within residential areas of cities (30 km/h). A small levy was imposed on heavy trucks on German motorways.

- Indirectly, the legislation on the recycling of packing materials, cars and electric/electronic appliances will have a substantial impact on the recycling rates of energy-intensive materials.

But since the last three years no measures, except investments in railways and the levy on trucks, have been taken up by the Federal Government. The Federal Government, however, seems be getting increasingly substantial support from public authorities' activities at a Länder and local government level. About half of the Länder Governments are considering adopting the 25 % reduction target in their own energy policy. Some 250 small and large cities have joined the "Climate Alliance", which has made a voluntary commitment to a 50 % reduction of urban CO2 emission by 2010. These Länder Governments and many city authorities are currently screening their possibilities for reduction by introducing local and regional energy concepts and by evaluating existing policies (see paragraph 2), although it seems to be rather impossible that the Climate Alliance Cities could meet their targets as long as general boundary conditions are not changed at the EU or federal level.

Concerned scientists and environmentalists are waiting to see how the German Federal Government will take up this issue and act in the next few years. A critical observation of trends will be very important as increasing per capita income and leisure time tend to support the energy-intensive consumption patterns of private households and substantially counteract improvements in energy efficiency (Schaefer, 1992). It is the view of the authors that the long term reduction of CO_2 in industrialised countries will go far beyond technical solutions in the energy and transportation sector, and that policy has to contribute to a changing value system and changing behaviour patterns associated with a less energy-intensive way of life. High unemployment, substantial financial transfers from West to East Germany in the order of 55,000 ECU per capita and year, a high migration surplus, particularly from eastern Europe and the former USSR, and short term operational interests of companies and voters will be the major arguments used to suggest that the 25 % reduction target is not achievable. In spite of all these difficulties there is still a possibility that activities at all levels of Government will find a way out of this dead-end situation in the interests of a sustainable earth.

References

Arbeitsgemeinschaft Energiebilanzen Energy Balance of the FR Germany, Frankfurt/Main, various years

Brand, M. et al.: Einfluß der Preisgestaltung leitungsgebundener Energieträger auf die rationelle Energieverwendung in Hessen, Wiesbaden, 1988

Brand, M., Jochem, E. "The Internal Energy Market: the New Coalition Against Energy Efficiency and Environmental Concerns?", Energy Policy, No 8, October 1990

BMU (Fed. Environment Ministry) Climate Protection in Germany. National Report of the FRG in anticipation of article 12 of the UN Framework Convention on Climate Change, Bonn, 1994.

Bundesministerium für Wirtschaft (BMWi) Energiepolitik in der Bundesrepublik Deutschland. Data on the development of the German energy economy (in German)in 1989, Bonn, 1990

BMWi Energiedaten '94, Bonn, 1994

BDI (Bundesverband der Deutschen Industrie) BDI-Bericht 1990-92, Köln, 1992, pp 113-125

BDI (Bundesverband der Deutschen Industrie) Erklärung der deutschen Wirtschaft zur Klimavorsorge, Köln, March 10, 1995

Clausnitzer, K.-D.; Hille, M Bestandsaufnahme der Arbeit von Energieagenturen. Inst. f. kommunale Energiewirtschaft und -politik, Bremen, Nov. 1993

CEU (Commission of the European Union, DG XVII) A View to the Future. Energy in Europe. Special Issue, Brussels, Sept. 1992

Climate Alliance Climate Alliance of European Cities with Indigenous Rainforest Peoples for the Protection of the Earth's Atmosphere, European Coordination Office, Frankfurt, Dec. 1992

Decker, M., P. Faross Die Konsequenzen der vorgeschlagenen Kohlenstoff-/Energiesteuer für den Energiesektor. Energy in Europe. 21 July 1993, p. 157-166

Enquête-Commission on Preventive Measures to Protect the Earth's Atmosphere: International Convention for the Protection of the Earth's Atmosphere, Avoidance and Reduction of Energy-Related Greenhouse Gases, 10 Volumes, Economica, Bonn, 1990a.

Enquête-Commission on Preventive Measures to Protect the Earth's Atmosphere Protecting the Earth: A Status Report with Recommendations for a New Energy Policy, Bonn, October 1990b.

Enquête-Commission on Preventive Measures to Protect the Earth's Atmosphere: Avoidance and Reduction of Energy-Related Greenhouse Gases, Volume 10, "Possibilities for Energy Policies and R&D Demand", Economica, Bonn, 1990c, p. 187.

Enquête Kommission „Schutz der Erdatmosphäre" (Hrsg.) Mehr Zukunft für die Erde. Economica, Bonn 1995.

Ewers, H.-J. Die monetären Schäden eines Super-GAU's in Biblis. Diskussionspapier Nr. 2 des Instituts für Verkehrswissenschaft an der Universität Münster, Münster 1991.

Federal Environment Ministry Environment Policy. Climate Protection in Germany, Bonn, 1993.

Federal Environment Ministry Synopsis of CO2 Reduction Measures and Potentials in Germany. Bonner Universitätsdruckerei, Bonn Dec. 1993.

Federal Environment Ministry Local Authority Climate Protection in the FRG, Bonn 1995.

Gruber, E. Efficiency of Energy Conservation Programmes in European Countries. Energy and Environment 3(1992)2, p. 122-132.

Gruber, E., Brand, M. "Promoting Energy Conservation in Small and Medium Sized Companies", Energy Policy, 19(1991)4, p. 279-287.

Heilemann, U.; B. Hillebrand "The German Coal Market after 1992", Energy Journal 13 (1992) 3, p. 141-156.

International Energy Agency (IEA) Energy Policies and Programmes of IEA Countries. 1989 Review. Paris 1990.

Ifo (Institut für Wirtschaftsforschung) Erfahrungen mit einsparpolitischen Maßnahmen in der Bundesrepublik Deutschland. in Enquête-Kommission Hrsg.: Energie und Klima. Bd. 10 Energiepolitische Handlungsmöglichkeiten und Forschungsbedarf. Economica Verlag Bonn, 1990. p. 482-503.

Jochem, E. "Long-Term Potentials of Rational Energy Use - The Unknown Possibilities for Reducing Greenhouse Gas Emissions", Energy and Environment, 2(1991)1, pp. 31-44.

Jochem, E.; Gruber, E. Obstacles to Rational Electricity Use and Measures to Alleviate them. Energy Policy 18(1990)5, p. 340-350.

Jochem, E; Schaefer, H. Emissionsminderung durch rationelle Energieverwendung. Energiewirtschaftliche Tagesfragen (1991)4, p. 207-215.

Moths, E. Die Preise schweigen, Merkur 530 (1993) p. 455-461.

Müller, E. How to solve inter-sectoral global problems? Difficulties of policy making - the case of CO2 reduction in Germany. German-Japanese Symposium "Global Environmental and Energy Policy after Rio". Manuscript Tokyo 17/18 Nov. 1992.

Prognos/IfE Die energiewirtschaftliche Entwicklung in der Bundesrepublik Deutschland bis zum Jahre 2010 unter Einbeziehung der fünf neuen Bundesländer. Basel 1991.

Prognos Identifizierung und Internalisierung der externen Kosten der Energieversorgung. Schäffer-Pöschel, Stuttgart 1993.

Schaefer, H. Aspekte und Möglichkeiten zum Energiehaushalten. BWK 44(1992)3, pp. 74-78.

Schiffer, H.-W. Energiemarkt in der Bundesrepublik Deutschland. TÜV Rheinland Verlag, Köln, various editions.

Statistisches Bundesamt Fachserie 18, Wiesbaden, several years.

UBA (Umweltbundesamt) Data of the Environment 1990-91 (in German), Berlin 1992, pp. 246-249.

Unger, H. Kernenergieerzeugung. BWK 44(1992)12, pp. 131-136.

VDEW Stand der Blockheizkraftwerkstechnik 1992. Ergebnisse der VDEW-Umfrage (H. Muders) Elektrizitätswirtschaft 92 (1993) 25, pp 1614-1618.

Wicke, L. Umweltökonomie. Eine praxisorientierte Einführung. 2. Ausgabe, München 1989.

4. Italian Energy Policy: From Planning to an (Imperfect) Market

Luigi De Paoli
Istituto di Economia delle Fonti di Energia (IEFE)
Bocconi University
Milan
Italy

4.1 Introduction

Energy has unquestionably been the subject of public intervention for a long time though its political importance dates from the first oil shock in 1973. This is particularly true in the case of Italy which, at the time depended on oil, almost all of it imported, for approximately 75% of its energy needs. Consequently, it was obvious that the risk of an energy supply shortage was a problem of paramount political importance. But with time, the attitude of Italian politicians towards this issue has changed quite substantially and the old concerns of energy policy have to some extent diminished in importance. This change corresponds with the more general evolution in the spirit of public intervention in the economy as well as the changed perceptions of the problems posed by the wider national and international events.

The aim of this chapter is to explain the shift from a first long period when energy policy was based on a plan to a second recent period more relying on market mechanism. The chapter is divided in three parts. We start by providing a picture of the evolution of energy trends. In the second part we analyse the "period of National Energy Plans" (PENs) starting with the oil shock till the end of the 80's. This part is divided in two: first of all we summarise the objectives of energy policy as they have been evolving during the five official PENs and then we compare the evolution of the facts with what the official energy policy envisaged. In the last part we look at the new approaches and priorities in energy policy in the 1990s. As part of this analysis, a short examination of the recent initiatives toward privatisation of energy sector will be made. Moreover we look briefly at the relationship between energy policy in Italy and in the European Union.

4.2 The Evolution of the Energy Situation in Italy

When the Yom Kippur war broke out, the Italian energy situation was characterised by a very high dependence on foreign supplies (upward of 80% of energy was imported), by a concentration in the use of oil (three quarters of energy consumption was oil-based) and by a rapidly increasing demand (the growth rate was 7-8% per year). The four-fold increase in the price of oil and the threat to this source's availability thus forced Italy to attempt a rapid change in its energy position. In effect, energy policy was one of the most important aspects of the Government's economic policy for a large part of the past twenty years. What have been the results of that effort?

The increase in demand for energy in Italy was very low between the first and second oil shock, it was negative between 1979 and 1983, and then recovered at an average rate of little more than 2% per year (see Table 4.1). In all, energy demand has grown by about 20% in 20 years, i.e. at an average rate of 1.0% per year, against 8.3% per year between 1960 and 1973. It is well known that this change in the energy growth rate, which has come about in all industrialised countries, has not only been due to the increase in the price of energy and the technical transformation this has brought about. At the same time as the energy crisis, the industrialised economies stagnated and it was not until 1984 that steady growth resumed.

Table 4.1:   Italy's Energy Requirements by Energy Source

mtoe	1973	1979	1985	1991	1993
Solid fuels	10.2	11.3	16.1	15.1	11.5
Natural gas	14.3	22.9	27.3	41.5	42.3
Oil	105.3	102.1	85.6	91.7	93.0
Hydro/geothermal	9.1	11.2	10.4	10.7	10.6
Nuclear power	0.7	0.5	1.5	0.0	0.0
Traded electricity	0.2	1.2	5.2	7.7	8.6
Total	139.8	149.2	146.0	166.7	166.0

Note: The energy equivalent of primary electricity is based on the fuel required to produce the same amount of electricity in thermal power stations.
Source : Ministry of Industry, Bilancio Energetico Nazionale, various years.

Energy intensity has decreased by about 25% during the period under consideration, placing Italy fully in line with trends in other industrialised countries.[1] It should, however, be pointed out that the drop in energy intensity took place primarily in the period 1976-1982. Since energy prices began falling in real terms, energy intensity has remained almost stationary. Moreover, it is well known that is not very useful to make comparisons in the time and space frame at an aggregate level because this hides structural differences and variations which can explain the evolution in energy demand. In Table 2 the final energy consumption is disaggregated into the three classical sectors (industry, transport, domestic and service industries) plus non-energy uses. As can be seen, the composition of final energy demand has changed significantly in Italy: the transport, service and domestic sectors have increased in importance in comparison to industry. Despite industrial production increasing by 40% in the last twenty years, during the same period energy consumption has dropped albeit slightly. In Italy, as elsewhere, industry is therefore mainly responsible for slowing down the growth in energy consumption, but, as in other economic activities, energy efficiency has almost ceased to rise after the mid-80's.

Table 4.2:   Italian Energy Consumption by Sector

mtoe	1973	1979	1985	1991	1993
Industry	50.4	51.3	44.0	50.6	48.4
Transport	27.4	30.7	32.1	38.9	40.6
Residential/Commercial	37.5	42.5	45.6	54.1	54.0
NonEnergy Uses	9.5	7.9	8.0	8.3	8.3
Total	124.8	132.4	129.7	151.9	151.3

Source: Ministry of Budget, Relazione generale sulla situazione economica del Paese, various.

---

1 Nevertheless, despite the "objectivity" which is at times attributed to the energy intensity indicator in international comparisons, one should not forget the numerous possible criticisms of the indicator. In particular those regarding: the inaccuracy in the GNP (and to a lesser degree the consumption of energy) of each country (eg. in Italy the estimated "black economy" led in 1987 to a substantial correction in the previous GNP evaluation); the distortion introduced by the transformation of the GNP into a common currency on the basis of exchange rates; the non-consideration of energy imported or exported indirectly through manufactured and semi-finished products.

As far as the structure of energy sources is concerned, the trend has been, as in other countries, to reduce oil consumption: oil's share decreased from 75% in 1973 to 56% in 1993 (see Table 1). This result was obtained by increasing the use of natural gas (from 10% to 25.5%) and net electricity imports (from 0.1% to 5.2%). [2]

In Italy, the rate of dependence on foreign supplies has remained almost constant, above 80%, due to the scarcity of traditional energy resources and the difficulty in developing the use of renewable sources which nevertheless, as will be seen, constitute one of the constant objectives of national energy policy. In economic terms, due to the large price movements, the impact of energy dependence on foreign supplies has undergone strong variations. The deficit of the balance of energy payments on GNP increased from less than 2% in 1973 to almost 7% in 1981 and then decreased progressively and went back to the same level of 1973. This was due above all to the link between the cost of Italian energy imports and the price of oil, the reduction in the energy intensity of GNP and in a very moderate degree, to the shifting of imports towards less expensive sources.

In summary, it can be said that, overall, changes in energy balances have been good in terms of containing demand growth (especially taking into account the low level of per capita energy consumption in Italy), modest in terms of reducing the importance of oil but insignificant in terms of enhancing energy independence. How can these results be assessed with respect to the objectives of energy policies and how does the evolution of these magnitudes differ from what would have taken place in the absence of public intervention?

4.3   Italian Energy Plans: Objectives and Instruments

Objectives

The Italian Government and Parliament have elaborated, discussed and approved five PENs: in 1975, 1977, 1981, 1985 and 1988. They were based upon a planning logic rather far removed from the traditions and above all the capacities of Italian public administration. There has been a progressive awareness of the incapacity to achieve what was indicated and of the limited political credibility of the PENs.

---

2 The role of electricity has been estimated on the basis of the quantity necessary for producing it with fossil fuels.

Nevertheless, the PEN has continued to serve as a reference text not so much for the quantitative objectives indicated therein as for the envisaged direction to take. Their content therefore remains a testimony to the evolution of Italian policy makers' perceptions of energy policy issues.

Looking at the fundamental purpose of the energy policy, the 1985 PEN expresses it as follows: "The main objective of the energy plan is to indicate the terms, procedures and actions to ensure that the country has the energy (in the short or longer term) necessary for sustaining economic, industrial and social growth at the lowest possible cost and with the maximum safety obtainable. Everything oriented to a progressive lessening of the restrictions imposed by energy supplies on the trade balance". In the 1988 PEN, the following was instead written: "The purpose of the energy policy is to ensure that the country has the necessary amount of energy available qualitatively and quantitatively. All this with respect for the environment and with final price conditions on a par with international competition, ensuring the stability of prices as far as possible".

As can be seen, the fundamental objectives have not changed from one Plan to the other nor with respect to previous Plans. They remain the guaranteeing of energy availability at the minimum cost. However there have been some adaptations which reflect particular preoccupations and which link energy policy with broader social and economic policies. For example, in the 1985 PEN, agreed at the end of a long period of economic crisis, it is underlined that the energy policy also had to reduce external constraints and encourage full employment. In the 1988 PEN on the other hand, when concern for the environment became increasingly strong in Italy, there is an emphasis on the environmental aspects of energy policy.

The general objective to guarantee energy supplies at the lowest cost has been specified by more detailed indications, among which the following are traditionally included:

- support for energy saving;
- development of national sources (renewable ones in particular);
- diversification of imported sources;
- geographical and political diversification of the supply areas.

The first objective has always been rather vaguely quantified due to the difficulty in establishing what is "spontaneous" and what is "added" in the natural trend towards slower consumption growth. It has often been quantified only in relation to the coefficient of elasticity in energy consumption to income, supposing that there were a "spontaneous" elasticity as well as a lower one resulting from the savings policy. In the 1981 PEN, for example, the savings objective was indicated as 15-20 Mtoe/year in 1990, which corresponded to an elasticity coefficient of 0.67. On the contrary, in 1988 PEN the saving objective of 10 Mtoe/year for the year 2000 is indicated which corresponded to an elasticity coefficient of 0.5. As a comparison, it should be borne in mind that the elasticity coefficient during the 1980-1990 period was 0.46. Can it therefore be said that the 1981 PEN energy saving objective has been overreached?

The objective of increasing the role of renewables has been accentuated over time. In the latest PEN, it is foreseen that the contribution of renewable sources would rise from 11 Mtoe in 1989 to 17 Mtoe in the year 2000. The increase, which is unlikely to be achieved, focused on the sources traditionally developed in Italy: hydroelectricity and geothermal energy. Yet potential for hydroelectricity has already been largely exploited. The increases which are still possible derive from many small plants with very high costs in most cases. [3]

The diversification objective has undergone numerous changes in the different Plans but it has always rotated around two main axes:: increased recourse to natural gas, especially for heating uses, and a reduction in the amount of oil, above all for electricity production. The main change has been in the means indicated to replace oil in the electricity sector. It is sufficient here to indicate that, in the latest PEN, the expected nuclear contribution has been eliminated whereas the natural gas contribution should rise from 21% in 1987 (25% in 1992) to 28% in 2000 (see Table 4.3), thanks above all to its role in electricity production.

The diversification of the areas where energy is imported from is an objective which is easier set than done. In fact the various energy plans have always avoided

---

3 At the National Conference on Energy in February 1987, ENEL (National Electricity Board) provided the following information: present hydroelectric capability: 45.7 TWh; capability reachable in 2000: 53.3 TWh. This required the building of 517 plants and the outlay of Lit. 9,900 billion.

providing precise indications on the subject. To illustrate what has been done, one can compare the 1970-1972 period - when Italian imports of hydrocarbons originated 61% from the Middle East, 27% from Algeria and Libya and 7% from the Soviet Union - and the 1988-1990 period when hydrocarbon imports originated in the amount of 26% from the Middle East, 30% from the two North African countries and 25.5% from the USSR. In other words, apart from a moderate reduction in the amounts from the three areas (from 95 to 82%), there has been an exchange of quotas between the Middle East and the USSR, from which Italy imports not only crude oil but also semi-finished products and natural gas. As Italy imports almost 60% of its hydrocarbons from only three countries (USSR, Libya and Algeria) none of which are very stable, it is hard to judge whether this policy guideline has been followed.

The objectives mentioned above do not completely encompass energy policy. In fact, if one accepts that the purpose of the energy policy is social well-being to be achieved by means of optimal allocation of resources (Le Bel, 1982), the policy should seek to correct all the violations to the conditions of free competition, particularly questions, such as externalities and market structures which have been rather neglected over much of this period.

Table 4.3: Italian Energy Requirements by Source According to 1988 PEN

mtoe	1987	1995	2000
Oil	90.0	87.5	81.0
Gas	32.3	42.0	50.0
Solid fuels	14.5	22.5	29.0
Hydro-Geothermal	10.1	12.5	14.0
Electricity imports	5.1	4.0	3.0
Other	1.0	4.0	3.0
Total	153.0	170.0	180.0
Energy import dependence (%)	81.0	79.0	76.0
Hydrocarbon import dependence (%)	69.0	64.0	59.0

Source: PEN 1988

Instruments and Institutions

The energy policy instruments which the Government can use to influence the behaviour of producers and consumers comprise: prices, which in this sector are strongly taxed when they are not directly set by the public authority; regulations (obligations or restrictions) established by law or by administrative measures; incentives (fiscal or financial) which can be granted to producers and consumers; and, information. In addition to these traditional and universal instruments (Maillet, 1980), another is of paramount importance in Italy: the guidelines given to the public sector.

In the case of Italy, the magnitude of the operational tasks of all kinds entrusted to State-owned firms has been such that one could think that the State should have been able, quickly and easily, to carry out the energy programmes it had itself established. This manner of reasoning implicitly assumes that politicians (Government and Parliament) had established not only general objectives (eg, to reduce dependence on oil), but also the method for attaining them (eg, by building a certain number of nuclear power plants) and the managers of firms (particularly ENEL and ENI) should have brought them about. Otherwise, the politicians would have sacked the managers.

An examination of the Italian situation shows that the reality is far removed from this interpretation for a number of reasons.[4] First of all, the process of the formation of objectives has, at best, seen some co-operation between politicians and managers and, at worst, domination on the part of the latter. In the second place, to consider politicians as perfectly disinterested parties pursuing nothing but the common interest would not explain all the political interventions that, in the energy field as in so many others, have ended up by distorting the attainment of officially stated objectives. Lastly, even supposing the politicians to be interested solely in the attainment of established energy policy objectives, the well-known problem would still exist regarding the asymmetrical information of controllers and controlled that

---

4 In actual fact, while the problem does have specific features in each domestic context, it is not merely Italian. See, for example, Finon, 1989 (Especially Introduction and Chapter 9).

has enabled State-owned firms to get out from under greater political control. [5] As these themes recur in any discussion of Italian energy policy, it is necessary to look briefly at the Italian institutional situation (even though the sector is undergoing reform).

In Italy, there has never been a Ministry of Energy, even if there have been frequent long discussions concerning its creation or that of a similar organism. The following two Ministries were specifically in charge of the sector and in particular were responsible for tutoring important public bodies or companies operating in the energy sector: the Ministry of Industry and the Ministry of State Participation (Ministero delle Partecipazioni Statali) which obviously operated in connection with Government decisions.

The Ministry of Industry is mainly in charge of government action in the field of energy. The body which, until its abolition at the end of 1994, set or supervised the majority of energy prices, the Interministerial Price Committee (CIP), was chaired by the Minister of Industry delegated by the Prime Minister. Many administrative procedures regarding the sector depend on this Ministry and above all ENEL and ENEA. ENEL (National Electricity Board), created in 1962, till now has monopolised the production, importation, transport and distribution of electricity, barring a few exceptions expressly foreseen by law. Given the importance of electricity in the energy system and in energy policy, control of ENEL is the element which placed the Ministry of Industry at the centre of national energy policy-making. ENEA (until 1991 the National Committee for Research and Development of Nuclear Energy and Alternative Energy) is the public body for research, promotion and development, at one time only in the nuclear field and subsequently also for other sources and energy saving. ENEA has also operational technical capacities in the environment field. In 1991 ENEA has had its scope modified by including the promotion of technical innovation. Its acronym has been kept, but the name now is "Agency for new technologies, energy and environment".

The Ministry of State Participation, made to measure for representing the interests of ENI (National Hydrocarbon Corporation) within the government at the time of

---

5 These themes are clearly an echo of more general problems of regulation, that is, of the State's intervention in economic activities other than through State-owned firms. Among the many reviews on this subject, see Peltzmann, 1989 and Noll, 1989.

Enrico Mattei, was abolished by a referendum in April 1993 and its duties have been subsequently transferred to the Industry Ministry. Up to 1993 the Ministry of State Participation had the task of supervising all public holdings in economic activity. The Ministry controlled three large bodies in which State participation were grouped together: IRI (Institute for Industrial Reconstruction), a large industrial holding also present in the financial sector, ENI which operates above all in the energy and chemical field and EFIM (Investment and Finance Board for Manufacturing Industries), the little brother of IRI, placed in liquidation in 1992 as a result of serious financial problems. These bodies, in turn, possess in whole or in part the share capital of operative private limited companies.

ENI was established in 1953 with the purpose of promoting initiatives of national interest in the field of hydrocarbons and chemicals. Hydrocarbons research activity, first in Italy and then abroad, has developed rapidly thanks to a high level of investments and to a collaboration policy with the producing countries. ENI has progressively extended its activity to the sectors of refining, transport and distribution of oil products and natural gas, becoming by far the leading Italian operator in all these fields. The activity of the Body was subsequently diversified into many other sectors more or less connected with the energy one. At the beginning of the nineties, ENI had thirteen sector-leading companies, four of which in the energy sector. AGIP dealt with exploration, production of hydrocarbons and supplies of crude oil in addition to what remains of the ENI commitment in the cycle of nuclear fuel and renewable sources. AGIP Petroli was responsible for refining and distribution of oil products. SNAM supervised supply, transport and sale to important purchasers of natural gas and controls the main gas distribution company (Italgas). Finally AGIP Coal was responsible for the entire coal cycle: research and mining, logistics, transformation and marketing on an international scale.

The presence of IRI is far more limited but not entirely negligible in the energy sector through industries which provide industrial goods for energy production. In particular, IRI controls, by means of the Finmeccanica holding company, Ansaldo which is the main Italian producer of plants for the production of electricity.

Thus the Italian energy sector has been and still is largely in the hands of State-owned companies, but, as noted, the situation is changing even though the final result is as yet unknown. In July 1992 the government announced its intention of privatising ENEL and ENI as well as IRI. In spite of several confirmations, however, this goal has not yet been achieved, mainly for reasons of finance and organisation, but also because of political opposition. Even so, there can be no doubt that, for the future, State-owned companies shall constitute a decreasing share of the instruments available for the execution of energy policies.

4.4 Objectives and Outcomes in Italian Energy Policy

Control of Demand: Prices, Incentives and Little Else

Energy demand has grown throughout the whole period in question at a much slower rate than Gross Domestic Product. Between 1979 and 1983, it even decreased continuously. The reduction in the growth rate has been much greater than the forecasts and the objectives of the various PENs, as we said. What has been done to obtain this result?

Prices have undoubtedly played a dominant role in limiting energy consumption of households and industry. The policy followed was that of maintaining high internal prices by means of a tax burden which was very high on average yet highly differentiated as to products and uses. For a long time, the use of such an important instrument has however been poorly analysed in the PENs. In the 1981 PEN there is only an indication that prices should be fixed on the basis of a "long term replacement costs" criterion. In the last two PENs, on the other hand, more detailed indications are provided. These include that the price paid by each consumer category must cover costs, that prices should tend to harmonise with those of the European Community and that the tax system should ensure neutrality of choice between the various sources "unless there is some specific determination aimed at achieving previously chosen energy objectives".

These indications have not however been transformed into facts yet, as the 1988 PEN points out a need to proceed gradually and bear in mind macro-economic restrictions. As a matter of fact tax policy has continued to be based on the following criteria:

- the guaranteeing of a substantial State revenue given the fact that the problem of covering the State Budget has become increasingly more difficult to tackle;

- the defence of a minimum level of consumption of families for social reasons;

- the modulation of tax increases, bearing in mind the consumption and products which are included in indexes to which wages are linked, in order to try and limit the inflationary effect.

Tax revenue linked to energy products (including the annual tax on cars) currently constitutes more than 15% of State tax revenue. Total revenue has constantly increased in real terms even if irregularly. In fact, after the oil shocks, the tax burden decreased to counteract the inflationary effect of increased prices, with a tendency towards recovery in successive years. On the other hand, after the oil counter-shock at the end of 1985, the decrease in price was almost entirely compensated by increased taxes. In this way an average level of taxation of oil products was achieved as never before and was probably unequalled in other industrialised countries. In fact, in recent years, the level of taxes has been from 2 to 3 times the cost of imported oil (Clo, 1991). The high tax burden has above all weighed heavily on oil products and, among these, on petrol (see Table 4.4). Strong imbalances in the prices of replaceable products has resulted, with consequent shifts in demand from one source or product to another.

Beginning in 1982, that is when gas began to be imported from Algeria, natural gas for heating purposes was taxed at a much lower rate than Diesel fuel in order to encourage substitution between the two products. Thanks to different tax rates, natural gas for heating cost about 75% of Diesel fuel in 1988 while still providing the gas distributor with a revenue that was 30-40% higher for equal energy contents (Ascari 1994). In accordance with the PEN's indications it was decided in 1988 that this advantage was no longer justified and tax on natural gas was increased substantially. But since in the meantime tax on Diesel fuel was also raised to ensure a revenue increase, the difference over the price of Diesel fuel was reduced but marginally, so much so that at the end of 1993 Diesel fuel for heating still cost some 20% more than natural gas (see Table 4.4).

The second case regards private transport. Diesel oil has been taxed much less than petrol. This was due to the inability of the administration to prevent a lower tax on gasoil for heating or for transport purposes from encouraging fraudulent behaviour.

It was thus felt preferable to set the price of Diesel fuel at a single level substantially lower than that of petrol. In order to compensate for this difference in price, a tax on diesel cars was introduced which did not however hinder their substantial diffusion in Italy for a certain period of time.[6]

Table 4.4   Energy Taxation in Italy: Petroleum Products

Lire per 10^7 calories	Gasoline	Gasoil (transport)	Gasoil (heating)	Natural Gas	Heavy Fuel Oil
1973	161.7	62.0	6.0	2.9	2.7
tax/price (%)	73.1	63.6	18.5	5.7	15.3
1975	264.1	75.4	13.2	3.4	6.9
tax/price (%)	66.6	46.2	14.5	5.7	12.1
1977	463.3	50.8	37.9	31.5	10.4
tax/price (%)	71.2	28.1	24.0	24.6	11.9
1979	482.3	57.6	51.8	49.6	13.6
tax/price (%)	69.3	23.6	23.1	26.0	11.5
1981	702.6	80.1	76.4	59.0	36.6
tax/price (%)	59.3	17.2	17.3	18.2	13.6
1983	976.6	196.9	188.6	69.5	51.5
tax/price (%)	63.8	28.7	28.9	14.9	15.8
1985	1 112.3	266.5	259.4	82.7	36.9
tax/price (%)	64.3	31.5	32.7	14.3	10.0
1987	1 334.0	437.9	429.1	72.8	24.5
tax/price (%)	78.6	57.0	60.6	14.2	13.9
1989	1 349.3	596.8	581.4	200.6	37.8
tax/price (%)	75.3	63.5	65.3	27.4	21.3
1991	1 503.0	949.1	912.3	384.2	111.2
tax/price (%)	75.9	70.3	69.7	40.7	43.5
1993	1 500.6	999.6	997.5	420.9	112.5
tax/price (%)	74.5	69.6	70.2	42.1	44.8

Source: Source: Ministry of Budget, Relazione generale sulla situazione economica del Paese, various.

---

6 The penetration of diesel cars had at first, however, reached saturation levels and then began to decrease, also because the ratio between petrol and diesel prices had decreased.

More recently some ministers and other representatives of the Administration have stated that the taxation of energy should be a function of explicit objectives for energy conservation and environmental protection. But many fear that this appeal may serve no purpose other than that of increasing the fiscal burden to raise revenues, thanks to the rigidity of energy demand. In a document dated November 1990, entitled "Piano Nazionale per il Risparmio di Energia" (National Plan for Energy Saving), the Ministry of Industry maintained that prices should be kept high to favour energy saving and in fact they should be further increased in order to finance, with additional tax revenue, the incentives to energy saving and use renewable sources. This document also proposed the setting up of "ecological taxes", a proposal supported both by the Minister of Industry as well as by the Minister of Environment, but opposed both by the industrialists and by trade unions. For the time being, these taxes have not been introduced in the same way as the principle has not been accepted that part of the energy taxation should be destined to financing the rational use of energy.

The burden of tax has represented the main mechanism for controlling energy demand. Other instruments have been used in a much more limited and belated manner. For example, the proposal to create a specific authority for the rational use of energy and the development of renewable sources was discussed at length in the late 70's when the problem of energy savings had an altogether different priority, but nothing came of it. On the other hand, in 1982, the CNEN (National Committee for Research in the Field of Nuclear Energy), was transformed into ENEA with the widening of its tasks to include also the rational use of energy and renewable sources. ENEA also serves as an instrument to aid the regions (which are by law responsible for many tasks even though they often do not have the necessary technical competence). These tasks were reconfirmed in 1991 when ENEA was reformed once again (Public Law No. 282/91).

The Government has been slow to devise incentives for energy saving. The law which provides for public energy saving support and the use of renewable sources was only approved in 1982 (Law 308/82) and it took 4-5 years to distribute the available funds (1,500 billion lire). This law was refinanced twice (in 1987 and 1989) until a new law, which sets out the regulations and incentives for the rational use of energy was approved in 1991 (Law 10/91). Due to budget difficulties, however, funds earmarked for the period 1991-93 (2,611 billion lire) were appropriated only to a very limited extent (some 400 billion). According to ENEA estimates (D'Angelo and Percuoco 1989), Law 308/82 has brought about investments for 5,800 b.lire in all, from which an energy saving of about 6 mtoe/year has been derived. Developing a policy for the rational use of energy has encountered many difficulties though it has permitted a saving which can be estimated at about 4% of current Italian consumption.

With regard to the use of standards, in 1976 Parliament approved the Law 373/76 which investigated more stringent criteria for the insulation of buildings and for the designing and operation of heating systems. Following on the second oil shock, the heating period (Law 178/80), the maximum temperature in buildings and the maximum speed of cars on roads and motorways were also limited by regulations. New rules for energy saving in the buildings trade have been introduced in 1991 (Law 10/91). Whereas regulations for designing buildings and heating systems can be considered as fairly effective, the measures regarding the behaviour of citizens in limiting heating consumption and transport are almost completely ineffective due to the incapacity and unwillingness of the Italian public administration to carry out controls.

Information as a means of re-orientating choices can have a substantial effect only if it encourages investments or influences the choice of equipment, given the fact that economies based on behaviour can be rapidly reversed. Both ENEL and ENEA have sought to diffuse knowledge regarding the possibility of energy saving but measures such as the indication of unitary consumption of pieces of equipment have never been imposed even if they are indicated in the PENs. Only in Law 10/91 is energy certification of buildings provided for (article 30), understood as supplying property buyers with supplementary information liable to influence price. The same law also introduced the requirement that firms with a yearly energy consumption of more than 10,000 toe employ an energy manager.

Diversification of Supply

As already stated, energy supply policy has as its purpose the guaranteeing of the availability and competitiveness of supplies above all by means of diversification which, for Italy, has first of all meant reducing the importance of oil.

The main substitute for oil in the heating market has been natural gas along with coal to a very small degree and solar energy to a negligible degree. The renewable source which has contributed the most has probably been wood, but official statistics are too haphazard on this point to be able to say so with any certainty. The spread of solar heaters has been encouraged by an ENEL campaign which guaranteed the installation of collectors with low-interest financing. ENEL in turn accepted an instalment payment with the electricity bills. However, when this campaign expired, it was not renewed. What is more, Law 308/82 for energy saving foresaw the possibility of obtaining subsidies for the installation of solar collectors. These campaigns have brought about an equivalent saving in primary sources estimated at 50-60,000 toe/year. This contribution is moderate, but it has allowed the solar collector market to develop, albeit at a very slow rate.

The other renewable, non-traditional sources of energy have developed even less. The use of low-temperature geothermal energy for urban heating, which at one time had aroused a great deal of interest, was not extensively applied in the country due to the difficulty of finding potential markets capable of sustaining its development under uncertain economic conditions. For the same reasons district heating, which had scarcely any role in Italy prior to the energy crisis, has been developed in 25 towns, but it covers only a very limited part of heating demand. At the end of 1993 the installed thermal power was slightly more than 1,800 thermal MW (di Marzio 1995).

The replacement of fuel oil with coal for thermal uses has come about almost exclusively in cement works. This change of fuel, corresponding to about 1 mtoe/year, is however very sensitive to price variations among coal, oil, coke, natural gas and fuel oil. It certainly cannot therefore be said that it is the result of energy policy measures.

A fundamental contribution to oil substitution has come from natural gas. Gas has become the "Italian way" to the (partial) substitution of oil and should continue to represent the natural alternative to it, above all in the case of the present diffusion programmes also including the electricity sector. Gas has also been the only source for which objectives have been fulfilled, in some cases with even better results than expected as shown in Table 4.5.

Table 4.5:   Projected Use of Natural Gas

Date of PEN	Horizon	Forecast Demand (bcm)	Actual Demand (bcm)	Forecast Share (%)	Actual Share (%)
1975	1985	500	33	15.6	18.7
1977	1985	42	33	16.8	18.7
1981	1990	42	47	18.9	24.1
1985	1995	40-44	>50	20.0	>25
1988	2000	60.6	?	28.0	?

Source: Italian Energy Plans

To understand the role that natural gas plays in Italy, one should bear in mind that it constitutes the country's major fossil fuel resource. In the 1950s, gas was already playing an important role in the industrial development of Northern Italy and allowed the accumulation of technical and management know-how in the transport company called SNAM. Mining revenue has also been the financial basis on which ENI has been able to build its expansion policy into the national market and to gain access to international resources.

The main reasons for the success of natural gas after the energy crisis are:

- managerial abilities at SNAM. This company has always been in a position to take prompt decisions in order to guarantee adequate supplies (even with decisions taken before the first oil shock) as well as to provide for the necessary infrastructures for transportation purposes;

- the choices of energy policy in favour of this source. Two were the major incentives granted for using this source: lower taxes than competitive products (see above) and subsidies to build urban district networks;

- lower environmental impact. This advantage has become of major importance over the last few years. Attention to this matter has increased enormously and social acceptance of huge energy projects has become a necessary condition to their feasibility.

As for the exploitation of gas in Italy from 1973 to the present day, three periods can be highlighted. Over the decade 1973-1983, the increase in availability due to gas imports from The Netherlands and the USSR was almost completely absorbed by the household and services sectors (see Table 4.6).

Table 4.6:   Natural Gas Consumption by Sector

b.cubic metres	1973	1977	1981	1985	1989	1993
Households	4.8	9.1	11.2	14.4	17.9	22.4
Industry	8.8	11.8	10.7	9.6	15.3	17.7
Transport	0.2	0.3	0.3	0.3	0.3	0.3
Chemical feedstock	2.2	2.0	2.1	2.4	2.3	1.1
Electric power production	1.2	3.0	2.3	6.3	8.4	9.7
Total	17.2	26.2	26.6	33.0	44.2	51.2
Share of final energy demand (%)	10.2	15.5	15.4	18.7	22.9	25.5

Note: Electricity production includes ENEL, municipal utilities and self producers
Source: Ministry of Industry, Bilancio Energetico Nazionale, various years.

During the second period (1983-1988), which was characterised by the supply of 10-12 billion cubic metres from Algeria and a second agreement with the USSR (see Table 4.7), placing the great quantities of in-coming natural gas was becoming somewhat more difficult. This was due to the necessity of building new pipeline networks in cities and above all to the fact that demand from industry proved to be lower than expected. It was not possible for ENI, in a short time, to sell all the additional gas in the sectors where it could be used at its best. Therefore ENI considered the possibility of selling a certain quantity of natural gas to the electricity sector. Beyond ideological polemics as to the real consistency of that behaviour (the use of gas in the production of electricity was considered in some places to be incorrect), this operation succeeded (see Table 4.6) as ENI sold natural gas to ENEL at a price which was equal to (and even a bit lower than) heavy fuel oil. Nevertheless, in the PEN updating of 1985, it was affirmed in any case that the use of gas in substantial quantities for energy production had to be considered as a transitory phase, to overcome the phase of excess of imported gas.

Table 4.7:   Production and Import of Natural Gas

b.cubic metres	1973	1977	1981	1985	1989	1993
Production	15.4	13.7	14	16.0	17.0	19.1
Total imports	2.0	12.9	13.9	19.3	28.1	32.1
- Libya	2.0	2.6	0.0	0.3	0.3	0.0
- Netherlands	0.0	3.6	6.5	4.6	5.6	5.4
- USSR	0.0	6.7	7.4	6.2	11.5	13.5
- Algeria	0.0	0.0	0.0	8.1	10.7	13.3

Source: Ministry of Industry, Bilancio Energetico Nazionale, various years.

The beginning of the third period coincides with the approval of the latest PEN. In fact, not only does the 1988 PEN no longer hint at the transitory use of natural gas in the electricity sector, but it also deletes the indication contained in the previous Plan as to limiting the importance of gas to 20% of total energy consumption. Indications contained in the Plan, in fact, show that gas consumption should increase to 51 b. cubic metres in 1995 and to approximately 60 b.cubic metres in 2000, taking penetration of this source to 25% and 28%, respectively. As a 20 b.cubic metres increase in consumption could not occur only in the household and industrial sectors, the objective of consuming 60 b.cubic metres in the year 2000 was therefore linked to the increase in gas use in the electricity sector.

When the 1988 PEN was issued, the choice of gas appeared to be one of the most peculiar characteristics of the new energy policy. To many observers, the objectives seemed to be too ambitious, as had in the past seemed the development objectives of the nuclear and coal sources. The programmes that followed proved once more that gas is an exception in Italian energy policy. Not only has there been no discussion of the 60 b.cubic metres within 2000, but not even of the new electricity programmes of ENEL's, largely based on gas which boosted the forecasted consumption of gas to 75 or even 80 b.cubic metres within 2000.

Considering the role that natural gas has played, it might be asked how the role of gas has been rationalised within the PENs? In fact, gas has never been considered, not even in the latest PEN, as a fundamental option for oil substitution in Italy. Perhaps this is due to the fact that no one wanted to state that gas has a sure advantage, current and future, over oil as to guarantee of supply and/or price.

The various Plans therefore merely indicated quantities of gas to be used at the various dates. As long as the share of gas was due to increase moderately, not much attention was paid to this fuel. In reality, it was SNAM that decided the sales targets for gas and the supply policy to guarantee the fulfilment of demand. In the 80's it was clear that gas had to replace oil more than to the expected extent, in order to reach the level of planned sales. For this purpose, a balance had to be struck within ENI between gas interests and the Group's general interests and secondly between ENI and the other operators.

It can, however, be said that overall, gas policy objectives have been proposed by ENI and have been totally or partially accepted by the government itself. This is but one of the examples of a more general rule: public enterprises in certain sectors are not merely the executors of political directions but become a source of strategies later adopted by the government. This may be considered a common and understandable procedure as these enterprises and bodies have the most appropriate technical know-how and are always aware of any possible problems that may arise in their sector. There is however a risk that these enterprises, often operating in a monopoly regime, may pursue their development and power more easily than a private company, in the belief that the public interest and that of State-owned enterprises are one and the same.

There have been instances where supply policies were subjected to heavy pressure by the government, for example the supply contract with Algeria (when the Foreign Trade Ministry imposed on SNAM the acceptance of a much higher price than the company was willing to pay) and the second import contract with the USSR (when the Italian Government delayed final agreement). However, as a rule, it is ENI that has decided the overall orientation of supply strategy and it has done so on the basis of diversifying its supplier base as much for its own ends as for the country as a whole.

Policies for the Electricity Sector

The electricity sector has always been central to Italian energy policy and particularly in connection with attempts to reduce the share of oil in the energy balance. Since the 1975 PEN, forecasts have always indicated (except for the 1977 PEN) a decrease in oil share of about 15% over a period of 10 years and a comparable share increase of sources other than oil in the production of electric power.

The first two Plans were based only on nuclear power for the refurbishing of the electricity generating industry. In reality, between 1973 and 1980, orders of thermoelectric plants based on fuel oil greatly exceeded those of nuclear plants. The 1981 and 1985 Plans were based instead on a double option: nuclear power was still the basic choice in the long term, but this was backed up by an important programme of coal-fired power plants to be built in a short time in order to accelerate fuel oil substitution. The fact that nuclear power incidence over total forecast consumption be kept steady at about 5% from the 1977 PEN through to the 1985 PEN, in spite of a longer time horizon, indicates that programmes have been regularly frozen and delayed for ten years. The three referenda in November 1987 on nuclear power, and the positions taken by Italian political parties,[7] led to the writing off of nuclear power from the theoretically feasible options. A year later, the Government decided to shut down all operating nuclear power plant and to reopen the nuclear option only when it would be possible to build "inherently safe" reactors.

Much of what has been stated for nuclear power plants is also applicable to the programme regarding coal-fired power plants. The increase in the use of coal was due to the conversion of oil-fired power plants between 1978 and 1985. From that date on, consumption remained at about 10 million tons. None of the few new coal-fired plants is still using this fuel. Most of them have not been completed but the use of coal is not foreseen in their initial phase of operation. From the mid-80's on, with the slump in oil price and the increasing concern of public opinion and politicians as

---

7   In Italy, referenda can only abrogate existing laws. To promote any change, citizens have to be called on to abolish a law or an article referring to that issue. 500,000 signatures then have to be collected for the referendum to take place. Proposition referenda are not provided for. This procedure favours ambiguous interpretations of questions and results of a referendum. As to the issue of nuclear power, for example, citizens were not asked whether they wanted nuclear energy or not, but whether they wished the abrogation of a number of indications related to those of nuclear energy.

to environmental problems, ENEL reviewed the coal option in an option in favour of multi-fuel power plants.

Electricity imports also had a rapid growth towards the mid-80's as electric power could be bought from France (either directly or indirectly through Switzerland) at competitive prices. At first, this condition was considered as temporary, but imports were maintained at the same, and latterly even higher, levels. In the end, in 1989, Italy had to import electricity to meet home demand for the first time. This structural dependence on electricity imports, even though it runs counter to all statements of energy policy, was implicitly acknowledged in the 1985 PEN which forecast that, in 1995, imports of electricity would account for 8% of total needs (compared with 10% in 1985). A similar implicit acknowledgement is present in the 1988 PEN, though this also foresaw a reduction in imports (from 11% in 1987 to 4% in 2000); in fact, imports have so far continued to increase (15.9% in 1993), a dependence which is officially accepted by the government.

The 1988 PEN forecasts indicated that the search for cost effective alternatives to the use of hydrocarbons had been effectively abandoned, largely as a result of decisions within ENEL. Who decided these programmes and their modifications? ENEL, holding the public monopoly for the generation and distribution of electricity, contributed to a remarkable extent to setting the agenda, albeit certainly less unilaterally than ENI did in the gas sector. This for at least three reasons:

- ENEL, as the holder of the generating monopoly, was certain to cover costs but had no profit motives. Consequently the company was largely indifferent to the choice of fuels;
- ENEL's monopoly status also blunted its incentives to extend the market electricity (for example in heating) by making its product more competitive with other sources.
- ENEL's programmes had to receive prior approval by a Ministry Committee (CIPE). Indeed, due to public concerns, throughout ENEL's history, politicians have been involved in planning decisions both for nuclear power, and, to a lesser extent, for coal-fired power plants.

Like other utilities ENEL was responsible for optimistic forecasts which systematically overestimated growth rates for electricity demand. Errors were partly due to an overestimate of the elasticity coefficient and partly to mistaken evaluations of economic growth (which ENEL as a public body has always taken from official data). The gap between programmes and actual accomplishments in the electricity sector has been enormous. The reason why this happened is complex, but some elements seem to be unquestionable.

The first problem that prevented the accomplishment of the electricity investment programmes has been that of the location of plants. The power granted to local Municipalities and Regions as regards the location of plants required a strong political will on the part of the government in order to keep to time schedules and maintain the momentum of the programme. This was not the case in Italy, partly due to differences (real or tactical) between political parties on basic choices and partly due to internal divisions within the same parties where local representatives were often defending opposite positions to those taken at national level.

To overcome these obstacles, the government adopted a variety of legislative solutions. Initially these were rather authoritarian (allowing the government to proceed with construction in the absence of local approval)[8], but the lack of success of this approach led the government to adopt a more promotional strategy, guaranteeing funds to local communities depending upon the type of plant and proportional to the capacity installed and energy produced. These solutions, however, did not succeed either in forcing or in convincing the local population to accept power plants (not just nuclear but also coal-fired investments). Both the authoritarian and the promotional attempts have in been abandoned following referenda held in November 1987.[9] Subsequently, at the end of 1988, a new procedure for authorisation was approved. According to this procedure, the Ministry of Environment can take part in the selection of locations through the process of "environmental impact evaluation" (called VIA), compulsory for all electrical generating plants. It is too early to assess whether the procedure has been more effective than the others (and above all whether ENEL and the State-owned Administration have learnt to approach local communities in a better way) as, de facto, all programmes for new coal-fired plants have been put aside.

---

8 In Public Law 7/83 power granted to the Ministry of Industry by Law 393/75 was conferred to CIPE.

9 The third referendum had the aim to withdraw the authorization given to ENEL to participate in nuclear enterprises abroad.

Besides problems in siting new plants, the government's procedures for authorising and monitoring the construction and operation of plants have been a source of inefficiency and delay. These problems in turn have increased the construction costs of new plant and rendered ENEL's investment planning useless. They are symptomatic of a wider credibility problem for government interventions in the electricity sector. Public scepticism over the ability of the government to handle questions of environmental protection and safety (reinforced by disputes between different government agencies on such matters as nuclear safety[10]) have contributed to the problems of building new capacity.

It can therefore be said that the Italian State has not succeeded in carrying out electricity programmes, approved by Government and Parliament due to a lack of coherent political decisions and institutional shortcomings within the government bureaucracy. For its part, ENEL has obtained no better results in its sphere of competence. The fact that ENEL has had only a slender ambition of becoming the main player of energy policy is proved by a similar lack of entrepreneurial spirit in the setting up of power plants, i.e., in its relationships with manufacturers and the bureaucracy as well as in the fulfilment of the industrial role that ENEL claimed for itself.

4.5   A New Agenda for Italian Energy Policy?

From a Decisionist State to a Regulatory State

For some twenty years Italian energy policy was based on the twin assumption that energy supplies, and in particular that of oil, were a cause for concern and that direct intervention on the part of the State was an effective and efficient way of tackling this problem. Energy policy was dirigiste in nature, i.e., it aimed to establish exactly what it was necessary to do and it possessed the means to do so through the

---

[10] This was the case in a dispute between the DISP, the body in charge of technical authorizations and regulating nuclear power plants, and ISS (Higher Institute for Health), which is studying the consequences on health of emissions from power plants on behalf of the Ministry of Health.

compliance of the energy supply sector. [11].Events, however, have undermined the technical assumptions and ideological foundations of such a policy. It can in fact be said that, from the early 1990s, Italian energy policy began to be transformed from its dirigiste preoccupations to one that left scope for greater flexibility and freedom of initiative (within certain regulatory constraints). Of course, as energy policy is linked to economic policy, the change in the conception of the State's intervention in the economy is a lot more general than its mere energy application. The main signals of the new course can be identified in the creation of a competition authority in October 1990[12] and in the decision to privatise a large part of publicly-owned firms taken in July 1992.

While it is not altogether easy to identify what caused the traditional basis of energy policy to falter, it would seem that three events or sets of events were particularly important: the Gulf crisis, the crisis in planning and the crisis of State-owned firms as part of the more general financial crisis of the Italian State.

The development and the final result of the Gulf crisis have certainly had a direct implication for energy policy. If the outbreak of the crisis in August 1990, with the Iraqi invasion of Kuwait and the subsequent increase in oil prices, had relaunched fears of oil vulnerability and of the need to have an energy policy to tackle it, the quick conclusion of the war and the return to very low prices, the disappearance of Iraqi and Kuwaiti production notwithstanding, strengthened the conviction that the oil market could be looked at much like all other markets. On the one hand it was possible to look to the US government which could be relied upon to solve, by military means if need be, the thornier political questions, while, on the other, if even the disappearance from the market of two of the Gulf's main oil producers was unable to push prices up, it meant that fears of a physical supply disruption really were unfounded. The Gulf war therefore had an opposite effect to that of Yom Kippur: it provided reassurance regarding the availability and the price of oil.

---

11 We have already discussed the theme of whether it was the politicians or the managers of State-owned firms to decide the lines of action. The formal decision in any case was taken by the political powers.

12 But it should be pointed out that the creation of the Antitrust Authority in Italy arrived exactly one hundred years after the American Sherman Act, but also after the approval of the EC's antitrust law that remained under discussion for 17 years, no less, and that made further delay in the Italian legislative process almost inevitable.

The factors which undermined confidence in the role of planning in energy policy cannot be identified so clearly as the episode of the Gulf conflict. At least two elements may be mentioned, however, one of a general and international nature and the other more specific. First the crisis and then the ruinous fall of Eastern Europe's socialist regimes certainly undermined confidence in the planning potential of centralised structures. Closer to home in the energy field, the systematic non-achievement of the PEN's fundamental guidelines contributed to a growing scepticism about the role of political decisions in directing energy policy. These and other planning flops explain how the ideology of non-intervention, gaining ground as it was at the international level, spurred by positions taken by the Thatcher and Reagan administrations, ended up becoming prevalent in Italy as well.

The credibility and feasibility of traditional energy policy on the direct intervention by the State on the economy has also been undermined by a fact specific to Italy: the crisis of State-owned firms and of public-sector finances. To understand the importance of this factor, it should be considered that the Italian State controls not only a large part of the energy sector (through ENEL and ENI, as said earlier), and of services such as railways, the post office and telephones, but also a large part of the banking system, a few large insurance companies and a substantial share of medium- to large-size manufacturing industries.[13] In other terms, among all western countries, Italy was (and still is) the one with the most extensive share of State-owned firms. In only a few cases was such ownership justified on grounds of market failure (eg, in natural monopolies). In several others it has been the result of the steady absorption of privately-held firms in trouble or of the expansion of existing State-owned firms.

This bloating of the State-owned sector has become unacceptable for two interrelated reasons. Firstly, a number of State-owned firms had to rely continuously upon public funds both to sustain investments and to cover losses (which were in some cases substantial and almost systematic).[14] Secondly, State-owned firms

---

13 According to Mediobanca figures on the first 1,800 Italian industrial companies, billings of State-owned firms ranged over the past decade from 34 to 42% of the total. See Mediobanca, Dati cumulativi di 1807 società italiane (Cumulative Figures for 1,807 Italian Companies), Ed. 1993.

14 In 1992, out of the 265 State-owned firms in the Mediobanca sample, 139 were in the black and chalked up total profits of 2,334 billion lire, 126 were in the red for a total of

became a huge game preserve for political parties who appointed (or encouraged the appointment of) managers and hired manpower. The flow of favours naturally worked the other way as well: substantial sums of money flowed to parties and politicians, as the Italian judiciary has recently highlighted.

Overall, therefore, the Italian situation seems to confirm the thesis common to the Public Choice school (Crew and Rowley 1989) and to the Chicago school (Peltzmann 1989) that politicians prefer regulation (in this case under the form of State ownership) to any other instrument for the distribution of wealth, such as taxes or subsidies, as its consequences are far less transparent and thus escape the attention of those who pay this tax. But as they are mechanisms of redistribution, taxes and regulatory action under the form of State ownership risk becoming equivalent in the long run. When this takes place, and faced with the difficulties of using the transparent tax instrument, the politicians (however reluctantly) will give up State ownership as the instrument has become no less expensive to use politically than taxes.

The reasons leading to this conclusion with reference to the specific Italian situation are as follows. The State's deficit has reached alarming proportions and it must of necessity be contained as it is becoming increasingly difficult to finance. This could be attained by tightening fiscal pressure but this measure is less and less practicable as the pressure itself rises. Another way to help alleviate the problem of the State's deficit would be to privatise at least part of the economic sectors controlled by the State. Through privatisation it might be possible to reduce cross-subsidies between categories of users but especially external subsidies in favour of entire sectors that have been losing money structurally and are thus obliged to count on public funds. In the second place, through the sale of State-owned firms, the State receives funds that replace the need to collect more through taxes and decreases the need to disburse them as owner-investor in the sectors themselves.

It can therefore be said that factors both internal and international have caused a crisis in the traditional forms of energy policy and a push toward a more market oriented policy.

10,133 billion lire. See Mediobanca op. cit. p. XIX. During the past five years total losses amounted to 26,550 billion lire against profits of 14,868 billion lire.

Towards Deregulation of the Oil Sector

The oil sector in Italy has been subjected to rigid controls throughout the history of its development. Since 1927 the distribution of fuel has been declared to be the State's exclusive preserve (Cassese 1992), a right extended in 1934 to oil refining. Private operators have therefore been able to carry out these activities only on the basis of a concession issued with complete discretionary power by the competent State bodies. Their activities have also been subjected to a considerable series of rules and regulations that ranged from plant operation to the setting of prices.

The choice of placing the oil sector under close State control was determined by the peculiar historical period and by the political climate in which it was hatched but it was also used as time went by to defend different interests. In the postwar period, for instance, it was used to further the development of ENI (created in 1953) and, after the 1973 oil crisis, to attempt to secure the country's oil supplies. It has only been in the last few years that the rigid system of controls has begun to be dismantled. For example the concessionary regime for refining has been limited following legislation introduced in 1991. As a result a limited plant change requires a simple authorisation and not the stipulation of a new concession. (This law, however, is not yet fully operational as the relevant regulations have not been issued.)

But the most significant development which may lead to a deep transformation in the operation of the oil sector in Italy is the greater deregulation introduced in distribution activities, especially car fuel. This process has been accelerated by decisions of new competition (or Antitrust) authority. In 1993 it struck down an agreement among the oil companies, worked out with the Industry Ministry acting as go-between, to reduce the number of points of sale in the territory by 11% over three years. The objective of the agreement had been to improve the efficiency of the Italian distribution sector. This situation notwithstanding, no company was taking independent action as, given the fierce administrative constraints, an improvement in efficiency does not lead automatically to increases in sales and profits. The activity of distribution is in fact still labouring under the concessionary regime, new points of sale cannot be opened at will, opening hours are strictly regulated and product and services sold are limited by rigid government tables. The very margins of the operators have long been determined by the administration.

While fully recognising the existence of these constraints and that the reduction in the number of points of sale would have increased efficiency, the Antitrust authority struck down the agreement because it was a restraint of trade. In essence, the Antitrust authority requested that the Government remove all those constraints on distribution activities that were no longer justified by the defence of the public interest but which restricted the scope for independent initiative on the part of the companies. According to the Antitrust authority, it is only through an increased level of competition that it will be possible to arrive at a restructuring of the sector to further the interests of the consumer at large.

The Government seems to have taken up the invitation and decided on 30 September 1993 to put the deregulation of oil prices into practice. Up to that date oil prices had always been subjected to some form of State control, through a mixture of administrative and supervisory measures. Under the first regime, the CIP (the price control body) imposed a maximum selling price (and the tax share, separately), on the basis of criteria that went through several changes in time. Initially it was inspired by the principle of equality between costs and revenues (CIP 28 October 1977), then on the weekly average in five EC Countries (CIP 19 March 1980 and CIP 6 July 1982). Other minor corrections were introduced subsequently without any method being considered satisfactory to take the specific features of the Italian market into account. With the supervision regime, on the other hand, the companies set their own prices but they had to report them to the CIP's Secretariat. They received approval after checking that the level was in line with the established criteria, before they went into effect.

The system evolved over time. After the second world war an increasing number of oil products was subjected to administration until, by 1974, they were all included (CIP 26 June 1974). But between 1977, when the supervisory regime was introduced, and 1986, all products were progressively transferred to this regime, beginning with products of lesser importance or whose purchasers who could effectively compete independently, ending up with petrol where the seller's market power is substantial. [15]

---

15 The passage from one regime to the other and even the differences between the two have never been very clear. For example the CIP Measure dated 3 June 1986 made changes to the administration regime of some products, petrol and Diesel fuel among them.

Although the supervisory regime has been conceived right from its very start as a temporary measure in the transition to a complete price deregulation, it was necessary to wait until 1991 (CIP 31 July 1991) for the price of fuel oil and of other minor products to be deregulated. At the same time a lighter supervisory regime was introduced for oil used in motor vehicles and for heating purposes. With the new supervisory regime, reported prices went immediately into effect and CIP exercised only an "after-the-fact" control. This regime has also been abandoned now and, as already said, beginning in October 1993, oil prices in the Italian market have been deregulated.

It is thus possible to say that, just as the 1973 oil crisis lead to an increase in regulation for the oil sector, so the stability of international markets, a new political and administrative orientation and the actions of the Antitrust authority have led to a considerable reduction in administrative constraints. Obviously the freedom to set prices is not complete, especially when the tax burden raises industrial prices by three or four times. But the oil sector in Italy does represent the best example of the shift from a dirigiste policy based on direct controls to a policy oriented to the exploitation of the market's potential. This path has not been altogether completed and there remain some constraints to be removed. However, from now on, oil companies in Italy will not be able to complain so loudly about the State and they will have to tackle the development of competition, which was hitherto largely non-existent.

Privatisation and Regulation of Network Energy Systems

The electricity and natural gas sectors, dominated by ENEL and SNAM respectively, have ostensibly been the instruments for diversification of sources of supply in Italian energy policy. However, the results have been the opposite of those intended. They exceeded expectations in the gas sector and remained well below objectives in the electricity sector. Programmes for setting up new coal-fired and nuclear power stations (before the option was abandoned) have never been fully achieved and have

---

However, subsequent CIP Measures refer to it when speaking of "extension of the supervisory regime". Practice as it has been observed confirms that the indicated products have never been subjected to a supervisory regime, even though the existence of such an error proves that there was a great difference between the two regimes. On this subject, see Gatti 1990.

been subject to construction delays, largely as a result of ENEL's inefficiency and the weakness of the State Administration. More recently Italy has become structurally dependent on imports of electricity. Recently, however, the problem of sources of supply, which remains important in the case of both electricity and natural gas, has been largely overshadowed by that of reforming the ownership and structure of these industries.

The intention of privatising both ENEL and ENI (SNAM's corporate parent) was announced by the Italian Government in July 1992 and took everybody by surprise. Once the decision was taken, it had to be put into practice and, as is the case in all complex processes that can be subject to different forms of development, contending ideas and interests sought to shape privatisation.

Privatisation offered an opportunity to attain several objectives, including the reduction of the enlarged public deficit, the introduction of greater efficiency in the productive system, the elimination of the worst excesses of improper use of State-owned firms to the advantage of politicians and their parties and the encouragement of greater pluralism in the Italian economy. The third and fourth objectives are concerned with the transfer of property rights and constitute the political nutshell of the privatisation problem. We will not examine them here in a distinct manner, but we will limit ourselves to consider them as privatisation's "political objective". This can be pursued in a manner consistent with both the objective of maximising revenue and with that of enhancing productive efficiency (though equally one could argue that the different objectives are in conflict with one another).

The problem is especially delicate in sectors where there are potential market failures, such as is the case of electricity and natural gas. It is not enough to decide what part of the capital to transfer (minority, majority, all of it) and how to proceed with the sale (through direct accords with industrial groups, with the transfer of a many small quotas, etc.). It is not even enough to be concerned with the stability of management and of the ownership structure.[16] It is also necessary to establish whether and how to restructure these sectors, that is whether the firms are to be split up before giving them back so as to introduce the highest degree of competition in the sector. It is lastly also necessary to define new rules and to ensure their

---

16 For example, through the constitution of stable shareholder cores as has been done in France with the privatisations of the 1986/88 period.

implementation without falling back into the trap of political interference. It is appropriate to expect that the way to proceed in tackling these problems and the solution adopted for privatisation depend on the relative weights given to the objectives mentioned above.

If the Government's priority objective were that of accelerating the privatisation process and maximising revenues, it is probable that the greatest attention would be given to creating the conditions for placing on the market at least a part of the ENEL and SNAM shares. Neither of the two companies would be dismembered vertically or horizontally. In fact, especially in the electricity sector, a single quasi-monopolistic firm would be seen by investors as the best option, reducing the risk to profits. It would of course in any case be necessary to redefine relationships between ENEL as a private company, the State and the other electricity producers and distributors, but this could be achieved through a gradual process of reform.

If the main objective were that of eliminating as soon as possible the possibilities of political interference in the operation of the firms, it is quite probable that the first step would be that of creating an independent regulatory authority, endowed with technical competencies and discretionary powers, albeit subjected to rules predetermined by the politicians. After that, at least the majority - and ideally the totality - of shares could be transferred to private hands. If the choice were to be made not to split ENEL into several companies, it is difficult to think of a solution other than that of the public company. This solution would in any case give birth to a new centre of domestic economic power. But it might be asked how far this objective would increase economic diversity and how much it might risk political and regulatory capture.[17]

Finally, if the main objective were that of encouraging the highest level of efficiency of electricity supply and the defence of consumers, it would not be necessary to consider the restructuring of the sector by introducing the highest possible degree of competition. In fact, the only thing one can be certain of with a change in ownership is that the structure of incentives for management is altered, but this does not, per se, lead to greater efficiency (Vickers and Yarrow 1988). It is therefore necessary, side by side with a change of ownership, to consider the influence that a change in

---

17 In effect, after privatisations were announced, Premier Giuliano Amato spoke about an occasion for the enhancement of a greater pluralism of economic power.

structure and in the sector's operating rules can have as a whole. Only by introducing a greater degree of actual or potential competition would it be possible to push the electricity industry toward a higher level of internal efficiency. Of course, the advantages of this solution should be weighed against the disadvantages linked to the loss of vertical and horizontal integration that are surely important in this sector (de Paoli 1993). If the current structure of the industry were retained it would be necessary to have a body to control prices and the behaviour of firms and thereby increase the industry's efficiency.

In the years since the decision to privatise was announced, the position of the various governments that have been in power has changed three times. The Amato and Ciampi Governments (April 1992-April 1994) were in favour of a privatisation of ENEL as a unique company accompanied by the creation of an independent regulatory authority. This indication is now a precondition since a Law passed in July 1994 makes it necessary that the sale of state-owned utilities "are subordinate to the creation of independent bodies for tariff regulation and quality of service control". A text of law instituting these bodies (for energy, telecommunications and transport) has been discussed during many months but, at present, it is not still approved, delaying the privatisation of ENEL.

In the meantime the Berlusconi Government (May 1994-December 1994) announced in November 1994 its intention of splitting up ENEL in two parts: a distribution and transmission company and a production company. Subsequently the production company had to be split up further in order that no single generating company could control more than 50% of the Italian market. This position has been reversed by the Dini Government (in power as of Summer 1995) which has opted to privatise ENEL as it is, citing the difficulties of splitting up the company and the urgency of privatisation (though it is trying to liberalise the generation side of the market).

SNAM, by contrast, was never intended as an early candidate for privatisation. There was a debate over the desirability of privatising ENI as a whole versus privatising its component companies. The difficulties have been compounded by SNAM's position as the company which has been the main source of profitability in ENI, offsetting the losses of its chemicals division. Since in 1994 the international cycle of the chemical industry has been good, this problem is now less difficult to tackle. Suddenly, in the first half of 1995, it appeared that the privatisation of the ENI holding company could be done quite quickly before ENEL's privatisation, changing the sequence that was originally envisaged.

If the objectives-actions grid described above is correct, it can be deduced from the actions of the governments that, at least as regards the electricity sector, the first concern was the quick appropriation of privatisation revenues. The objective of reducing the State's direct intervention in the economy persists but it has greater support from public opinion than from politicians. In fact, some in parliament and in the Italian political parties, with whom the government in any case has to come to terms for parliamentary support, are not very enthusiastic about privatisation. As to the efficiency objective, it has hardly any space at all in governmental decisions as it is not backed by industrial, social or political subjects that champion it with any vigour. All this leads us to feel that it is unlikely that great changes will take place in the short term in the electricity sector and in that of natural gas.

In spite of the intention of privatising ENEL as a whole and the almost total confirmation in the concession draft of the ENEL's existing monopolistic status,[18] the structure of the electricity sector in Italy appears likely to undergo some changes, especially in power generation. Most transformations now under way began before the decision to privatise was taken and are due both to the redefinition of ENEL's strategy and to changes in regulatory outlook on the part of Government and Parliament.

By the early 1990s, ENEL's management estimated that the cause of past failures in setting up electric power plants was the opposition to new power plants and drew the conclusion that it was unrealistic to assume that the Government would provide more adequate support than it had done in the past. The company therefore decided to redefine its strategy in the following way:

- to choose sources according to social acceptance and not according to economic considerations, especially as, in the long run, this was shown to have been wrong in the past due to uncertainty in the cost of both plants and fuel prices;

18 ENEL as a private company must obtain from the State a concession for carrying out its activity. Such a document has already been drafted and changed several times, but it has not yet received final approval.

- to shift the core of the business from generation to distribution due to the choice indicated above as well as to the acknowledgement of the fact that dependence on other producers (domestic and foreign) cannot be eliminated;

- to retain some margins of flexibility in the supply of sources and electric power in order to strengthen its commercial importance in accordance with the previous choice;

- to take into account internal limits using external contributions whenever this could help to solve the problem of guaranteeing an appropriate electricity supply (for example by buying desulphurated fuels instead of making desulphurisers).

In a nutshell, these changes entailed a revised strategy for determining both the sources of electricity supply and relationships with producers. Beginning with the 1988 PEN, the use of natural gas in the generation of electricity was no longer considered as a temporary exception. But at first gas was to be used in multi-fuel power stations, as a means to overcome opposition to the use of other sources, or in gas turbine stations. Subsequently, however, ENEL accepted the technology of combined-cycle plants and redirected all of its programmes toward that type of plant. More recently, ENEL seems to have given up trying to find new sites and is aiming at the introduction of combined cycles in existing plants through the addition of extra equipment to increase their capacity. Such an attitude has been substantially accepted by the Government, although some concern has been voiced as to the availability of natural gas.

As regards relationships with other electricity producers, new factors have been introduced mainly through legislative actions approved even before the decision to privatise ENEL. An important law passed in early 1991 entitled "Measures for the execution of the PEN", offered the possibility that a huge independent generation of electricity could be developed in Italy. The freedom of production of independent producers has been expanded substantially beyond the 1962 nationalisation law (de Paoli 1991). Firstly, generating companies can generate enough to meet their own requirements on a group basis, with ENEL being committed to guarantee transport against payment. The excess of electric power produced must be sold to ENEL as ENEL's monopoly for the sale of electricity to third parties has been totally maintained. But - and this is the second extension - in the case of electricity produced by renewable, special sources (assimilated with renewables) or cogeneration plants, ENEL is bound to purchase this energy at a price set by the authorities. In applying this norm, the CIP Measure No. 6/92 has set fairly high prices that have encouraged a substantial supply of new plants destined to generate electricity solely for the purpose of selling it to ENEL.

According to current forecasts ENEL by 2000 could be purchasing energy generated by 6-8, 000 MW of plants owned by independent producers. Even though this diminishes the role played by ENEL as a producer, this is in line with its interests as it reduces the need for investment capital [19] and does not place its quasi-monopoly on sales under discussion. But it is legitimate to expect that, through the creation of a more diversified electricity supply, an internal electricity market will be progressively created in which final producers exert pressure to gain direct access to final users. Thus, although the privatisation of ENEL does not seem to have this among its objectives, partial deregulation of the market advanced by previous strategies will probably lead to a more openly competitive electrical industry.

Natural gas is unlikely to undergo the same degree of change as the electricity sector. SNAM has a de-facto but not a de-jure monopoly in the transmission and importation of gas. The distribution of gas, however, dominated as it is by Italgas (which in turn is controlled by SNAM) consists of about 800 companies of different types and sizes. There would not therefore seem to be a great deal of room for interventions aiming at changing the sector's regulatory picture, though there have been some changes as regards access to the gas network. SNAM has been obliged by a law introduced in 1991 to transport gas produced in Italy and used by producers or by companies belonging to the same industrial group or used for supplies to ENEL or municipal utilities. A complete opening up of transmission networks has never been seriously envisaged, however.

The most important consequence of these changes may be a substantial increase in gas consumption for the generation of electricity. After supporting the use of gas in the electricity sector in the mid-80's (see above), SNAM does not seem to appreciate the opening of this new market. In the past the sale of gas to the electricity sector

---

19 ENEL has a debt/capital ratio equal to 0.6, constituting a serious problem for its privatisation. The reduction of debt is thus a key objective for ENEL management.

has not been profitable at all and has in fact been subsidised especially by household users (Ascari 1994). The expansion of this use would make such subsidies entirely unsustainable but, above all, it would raise the overall cost of SNAM's gas supplies. In fact, current contracts are sufficient for meeting demand for household and industrial uses and a small part of the electricity-generating market. If the electrical sector's demand for gas rises, SNAM must procure new gas from far-away sources or through LNG, certainly more expensive, or allow others to develop a commercial activity, something it cannot be expected to view with any favour.

However, if the Government deems it appropriate not to put a brake on the use of gas for the electricity-generating sector for reasons of economy or of security of supplies, it may pressure SNAM to take a more active attitude in the development of supplies, perhaps giving it informal assurances on the maintenance of its current treatment. It can therefore be said that, as regards Italy in the 1990s, traditional energy policy priorities, concerned with guaranteeing domestic supplies, prevail more in the natural gas sector than in oil..

4.6   The EU's Impact on Italian Energy Policy

The EU's energy policy has followed a path similar to that of Italian energy policy. Initially it emphasised energy security and planning as a means of achieving that objective. This approach lasted until the late 1980s when the Commission published its report on "The Internal Energy Market". The earlier attempts to enhance energy independence were largely ineffective notwithstanding the effects of the 1970s oil crises. Instead, the problem of the constitution and operation of oil stocks was discussed and decided on at the International Energy Agency while the Commission's attempts to plan future use of energy (especially in the electricity sector) failed to secure a social consensus on new options (such as nuclear power) or to constrain national decision-makers.

Documents and directives approved or proposed in the wake of the Single Act have brought about a radical change in this situation. The Single Act aimed to relaunch the process of European integration through greater economic integration by market liberalisation. If the free trade inspiration of the policy merely reflected the priorities of the Treaty of Rome, the emphasis on competition and competition policy marked a new departure. While the promotion of free trade is a factor which affects the joint welfare of the countries involved, the promotion of competition can have effects that are purely internal to any one country. This change in emphasis was reflected in European energy policy initiatives.

European energy policy has attached only a limited importance to increasing trade in oil products between member states, provided there was no legal monopoly in distribution or refining. It was felt that freedom of trade was guaranteed to a sufficient degree by the lack of customs or legal barriers. From the Commission's perspective, the real obstacle to a single market were the differences in environmental rules and fiscal conditions between states. The attempt to reach a harmonisation of excise taxes on oil products has substantially failed, however. It follows that from the point of view of the final users, the single market for these products exists only to a very limited extent. In this sector, European energy policy has thus not constituted a constraint on domestic policies. In fact, Italy has continued to make use of the consumption of oil products as an important source of tax revenues.

European policy for the energy network industries (gas and electricity) has been altogether different. In this case intra-European trade is inevitably limited: in the case of natural gas, by the fact that no country, except The Netherlands, has sufficient reserves to export to any extent, and in the case of electricity by technical constraints. In spite of this, and invoking the need to guarantee the free circulation of goods, the Commission has insisted upon putting forward proposals to open up these markets. In this way, the Commission has attempted to have an effect on the internal organisation of the electricity and gas sectors rather than on trade, justifying its actions on the need to promote the interests of consumers (Commission of the European Communities, 1991).

These proposals, which have yet to be approved by member states, have received a mixed reception in Italy. The proposals to liberalise electricity and natural gas markets are at variance with the Italian government's plans for privatising ENEL and SNAM. The government was also opposed to the Commission's attempts to abolish gas and electricity trade monopolies and seems likely to transfer this right to a privately owned ENEL.. In other respects, however, Italian and EU policies have been moving in the same direction. Electricity generation in Italy is now more open than it has been in the past while the moves towards deregulation of the oil sector have opened up that industry to competition. As for trade in electricity, Italy has adopted a very pragmatic and open stance.

Moreover, the Commission's indication in 1990 that national supply security might be replaced by that of the Community is supported by Italian practice. Italy, by combining interest, need and belief, seems willing to accept a shift in competence on supply security from a national to an European level. No one in Italy now believes that a domestic energy policy can be conceived as a task of exclusive competence of local political power. But this does not in any way mean a lessening of the role of government in each member state. It is precisely because of the loss of some freedom in setting energy policy instruments that there has to be an enhancement of energy policy coordination at a European level. (IEFE 1992). The transfer of power from the domestic level to that of the Community finds its point of balance in the principle of "subsidiarity" contained in the Treaty signed in Maastricht on 7 February 1992. This exercise is not simple, however, and there is a risk that too much will be transferred to the European level, including actions which might be best carried out at a national or a local level (de Paoli and Vacca 1991).

Differences among the member states and between them and the Commission is inevitable. This was a less serious problem in the past as European energy policy was not seen as very important and was considered ineffective and not binding. But a more effective Community energy policy demands greater cohesion and participation; without this, the failure to bring each member country's interests and capabilities to the negotiating table would inevitably lead either to non-compliance with the decisions adopted or to delayed compliance.

4.7   Summary and Conclusions

During the 1970s and 80s, the energy sector was probably the sector in which planning techniques played the most significant role, with security concerns and expectations of resource depletion justifying this strong political intervention. Italy, however, has not had a long tradition in planning and lacked the bureaucratic capabilities to make such a system effective. Moreover, the Italian political system, normally based on bargaining and concerned with a short-term evaluation of problems (as with most political systems), proved unwilling to take important strategic decisions. Italian energy policy, which initially aimed to limit the use of hydrocarbons and to reduce energy costs, succeeded to a certain extent in achieving these objectives. Despite its problems, therefore Italian energy policy was not a failure. On the one hand, the policy may have encouraged the adaptation of the Italian economy to oil shocks through a greater opening towards exports in order to pay the energy bill (Bourgeois, Criqui and Percebois, 1986). On the other hand, it enabled the energy system to develop a greater flexibility compared to countries that seriously accepted the premise of a price of hydrocarbons which was rapidly and continuously increasing. By the mid-80's, the slump in oil prices had rendered energy dependence less of a concern and weakened the urgency of following guidelines which were designed to guarantee energy supplies. Moreover, the inefficiency of past planning increased doubts as to whether this was the correct way to follow.

The strengthening, at the international level, of more liberal views of economic policy favoured the discussion of real efficiency of public intervention. This new trend has also been enhanced by the new European energy policy striving to create a single market rather than to ensure, as in the past, energy independence. The security of supplies is surely still an important element, but the solution proposed to reach this is new. The Lubbers proposal, that triggered the initiative for a "European Energy Charter" represents one of these solutions (Commission of the European Communities 1991). The means to achieve that goal is that of cooperation and integration, rather than diversification, towards more secure sources (nuclear and coal) and energy saving.

During the early 90's, conditions thus obtained for a change in the direction of energy policy. The concern of guaranteeing the availability of energy made way for that of ensuring greater efficiency by reducing direct actions on the part of the State and increasing the recourse to the market. The most important examples of this new direction have, to date, been two: the decision to privatise ENEL and ENI and to deregulate the prices of oil products. The decision to privatise the two State-owned companies dominating the electricity and hydrocarbons sectors has not, however, been fully implemented yet. Moreover, on the basis of intentions expressed so far, these privatisations do not seem altogether satisfactory if more efficient behaviour and more competitive markets are to be encouraged. This cannot come as a surprise: political decisions, even when they are inspired by a market objective, are never based on the unadulterated search for efficiency.

Indeed, just as it was incorrect to think that energy policy could take the place of the market, so it would be equally wrong to believe that its sole purpose is that of allowing the market to function undisturbed. There is no doubt that, what we call the "result of market forces" depends to a substantial extent on the social and institutional context within which the subjects operate in the market. If the new Italian energy policy wants to be more effective than the prior one, it is necessary for the political class to learn as soon as possible what it means to carry out the role of regulator, directly and indirectly.

References

Ascari S., "Regulatory Capture and Natural Gas Pricing in Italy", ENER Bulletin, No. 13, 1994.

Bourgeois, B. Criqui, P and Percebois, J "Politiques énergétiques et adaptations au nouveau contexte économique", Revue de l'Énergie, n. 388, 1986.

Bouttes J.P.and Lederer P., "Towards a New Industrial Organization of the Electricity Sector in Europe", Paper presented at the Conference Organizing and Regulating Electric Systems in the Nineties, Paris, May 28-30, 1990.

Cassese, S "La regolamentazione dei servizi di pubblica utilità in Italia" (The regulation of public services in Italy), L'industria, No. 2, 1992.

Clò A., "Congiuntura e politica energetica in Italia nel 1990" (The Economic Climate and Energy Policy in Italy in 1990), Energia, n.1, 1991.

Commission of the European Communities, com 88238

Commission of the European Communities, European Energy Charter, Com (91) 1991

Commission of the European Communities com 91 548, 1992

Crew M.A., Rowley C.K., "Feasibility of Deregulation: a Public Choice Analysis", M.C. Crew (Ed.), Deregulation and Diversification of Utilities, Kluwer Academic Pu., Boston, 1989.)

D'Angelo E., Percuoco A., "Un primo bilancio dei risultati conseguiti con la Legge 308/82", (First results from Law 308/82), Energia e Innovazione, n. 6, June 1989.

de Paoli, L "La regolamentazione dell'industria elettrica nella Comunità Europea" (Regulation in the electrical industry in the European Community) in Regolamentazione e mercato unico dell'energia, F. Angeli, Milan, 1993

de Paoli, L "Organization and Regulation of the Italian Electric System", ENER Bulletin n. 9, 1991.

de Paoli, L and Vacca, S "Politica energetica e direttive della Cee" (Energy Policy and the EEC Directives), Economia delle Fonti di Energia, No. 45, 1991

Di Marzio, T "Il teleriscaldamento nella provincia di Brescia: uno studio di fattibilità", (Feasibility study on district heating in the Province of Brescia), Economia delle fonti di energia e dell'ambiente, n. 1, 1995.

Finon, D , L'échec des surgénérateurs: autopsie d'un grand programme, Presses Universitaires de Grenoble, 1989.

Gatti, G , "Come uscire dalla trappola della finta sorveglianza dei prezzi " (How to escape the trap of bogus price supervision), Staffetta Quotidiana Petrolifera, 12 November 1990.

IEFE, "Il ruolo del MercatoUnico Europeo dell'energia e l'industria italiana" (The Role of the European Single Market for Energy and Italian Industry), Economia delle Fonti Energia, No. 48, 1992.

Le Bel, P.G., Energy Economics and Technology, The John Hopkins University Press, Baltimore, 1982

Maillet P., "Calcul économique et stratégies énergétiques", Revue de l'Énergie, No 325, 1980.

Noll, R "Economic Perspectives of Regulation", pp. 1254-1287 in R. Schmalensee, R.D. Willig, Handbook of Industrial Organisation, . Elsevier, Amsterdam, 1989.

Peltzmann, S, "The Economic Theory of Regulation After a Decade of Deregulation", Brookings Papers on Economic Activity, 1989, pp.1-41

Vickers, J and Yarrow, G Privatisation: an Economic Analysis, MIT Press, Cambridge (Mass.), 1988.

5. Ideology and Expediency in British Energy Policy

Francis McGowan
School of European Studies
University of Sussex
Falmer, East Sussex
United Kingdom

5.1 Introduction

Energy policy in the UK has undergone some dramatic changes over the last fifteen years. More so than any other European country, the government has adopted the rhetoric of promoting market forces over state intervention in the energy sector and has followed this up in policy terms with significant initiatives in the fields of privatisation and liberalisation. The changes have been such as to lead some to allege that the UK no longer has an "energy policy" and that short-termist myopia has replaced the longer term strategic vision which characterised UK energy policy in the past (and which arguably persists in other countries). However, while the changes of the last decade and a half have been substantial, they should not be regarded as demonstrating an unequivocal commitment to the market or as marking a disengagement from political intervention and interference. While the government gives much greater emphasis to considerations of efficiency - as a result of its ideological preoccupation with the market - there is some evidence that expediency or pragmatism has often prevailed in the shaping of policies. If policy is myopic it stems as much from short term political calculations as from the market-place.

This change in policy has taken place against a dramatic transformation of UK energy balances. The embrace of the market coincided with the UK becoming self-sufficient in energy, enjoying a portfolio of energy resources unmatched anywhere in the European Community. Some have argued that these two changes are connected, that only a well-endowed country could afford the luxury of a market-driven energy strategy, downgrading the traditional concern of policy makers with supply security in favour of other objectives. Whether or not the policy shift was only practical in such a strong supply-side context, it is clear that the policy has begun to affect the shape of UK energy balances - and will do so in the future - although so far it has not undermined the government's overall approach to energy.

In this chapter we review the development of UK energy policy before and after 1979, the year when the first of a series of Conservative governments committed to liberalisation of the British economy was elected. It focuses on the policies of privatisation and liberalisation which have been developed over that period and consider the consequences of these changes, particularly in the electricity and gas sectors. In both sectors, the government has created independent regulatory bodies and the chapter assesses their impact on energy policy. The chapter also examines how the British government has reconciled its approach with other policy concerns, notably the environment. In this and in other respects, UK energy policy has been affected by developments in the European Union (EU) and the chapter reviews the ambivalent relationship between the UK and the EU on energy matters.

5.2   The Energy Sector in the UK

Before reviewing the development of British energy policy we outline the basic characteristics of the UK energy sector. Even before the North Sea discoveries, the UK was well-endowed with energy resources - historically it was a significant exporter of coal - and energy has always been important for the UK economy as a sector in its own right, supporting a range of engineering and supplier industries, the interests of which were often to the fore in energy policy-making.

Coal

The production and use of coal in the UK has been in more or less continuous decline over the post war period; from a 1950 output level of 216m. tonnes, production had fallen to 130m. tonnes by 1980 and to 48m. tonnes by 1994 (see Table 5.1). Demand followed a similar pattern while, in recent years, more requirements have been met from imported sources (between 15-20m.tonnes in the 1990s). The fortunes of the sector have been overseen by British Coal (formerly the National Coal Board), a nationalised industry established in 1947. It was responsible for all significant deep mined production (open-cast activities were often licensed to private contractors) and much of the marketing of coal. In 1995, coal returned to the private sector with the bulk (70%) of the slimmed down industry operated by RJB Mining (UK) Ltd.

Table 5.1:   British Coal Production and Trade

m.tonnes	1950	1960	1970	1980	1990	1994
Production	216.0	197.8	147.2	130.1	92.8	48.0
Trade	-17.1	-5.6	-3.1	3.5	12.5	14.8

Note: Negative values refer to net exports; positive values to net imports.

Source: UK Energy Digest, various years.

Oil

Oil requirements in the UK have grown from 15m.tonnes in 1950 to just over 71m.tonnes in 1994 while final use increased from 12.5m.tonnes to 66.2m.tonnes over the same period (see Table 5.2). Since the 1970s, much of the raw material has been produced in UK offshore waters and the country has been a net exporter of crude oil since 1980. Oil exploration and production was carried out by a mixture of public and privately owned companies though, after the privatisations of the 1980s, this is now a purely private activity. Downstream activities have been dominated by the private sector; while the government had a significant stake in British Petroleum for many years, this shareholding was gradually privatised.

Table 5.2:   Crude Oil Inland Consumption and Production in the UK

m.tonnes	1950	1960	1970	1980	1990	1994
Consumption	15.0	42.9	88.2	71.4	71.6	71.2
Production	0.2	0.1	0.2	80.5	91.6	126.7

Source: UK Energy Digest, various years.

Gas

Gas supply has increased from 58.9 Twh in 1950 when it was wholly manufactured to 760 Twh in 1994, when it was wholly natural gas (see Table 5.3). The development of natural gas from production to final supply was presided over by British Gas a single nationalised entity which replaced the patchwork of municipal and privately owned local utilities in the late 1940s. British Gas had substantial interests in gas (and oil) production (though other companies were also involved in exploration and extraction) and enjoyed for many years a monopsony position in gas purchasing and a monopoly in transmission and sales of gas. In 1981 these monopoly privileges were removed (see below) and in 1986 British Gas was privatised. Over the following decade, competition has increased in the gas market.

Table 5.3:   UK Gas Supply: Manufactured and Natural

TWh	1950	1960	1970	1980	1990	1994
Manufactured Gas	58.9	66.6	49.6	0.6	0	0
Natural Gas	0	0.8	121.7	404.8	531.4	760.3

Source: UK Energy Digest, various years.

Electricity

Electricity supply has increased from just under 70 Twh in 1950 to 308 Twh in 1994. Until recently the bulk of production needs were met by publicly owned suppliers while self-generation and private sales to the grid were in relative decline (see Table 5.4). As a nationalised industry, distribution of electricity in England and Wales was carried out by 12 Area Boards and production and transmission was focused in the Central Electricity Generating Board (supplies in Scotland and Northern Ireland were carried out by vertically integrated companies). Privatisation, beginning in 1990, has transformed the industry, breaking up the CEGB into separate production (3) and transmission companies in England and Wales while competition in production and the supply of final customers has also increased.

Table 5.4:   Electricity Production in the UK

TWh	1950	1960	1970	1980	1990	1994
Public Supply	55.8	113.4	212.4	249.1	280.6	286.9
Other	12.0	15.6	19.2	17.2	19.5	21.0

Source: UK Energy Digest, various years.

5.3   UK Energy Policy Before 1979

For much of the post-war period until the election of Margaret Thatcher in 1979, British energy policy was largely conducted through the publicly-owned firms which dominated the energy sector following the nationalisation of the coal, electricity and gas industries in the late 1940s (the government already maintained a stake in the oil industry through a shareholding in what was to become British Petroleum). Indeed, just as privatisation might be seen as the guiding principle of policy for most of the period since 1979, so in earlier times public ownership played a similar role: as the owner of the key energy industries, the state had a direct means of influencing policy. For most of the period, moreover, energy policy was largely bipartisan (so much so that there was little need to present an explicit strategy for the sector). Nationalisation of these industries was moderately controversial when it took place, but it was not long before the transfer of these industries into public hands was recognised as an accomplished fact.[1]

Prior to the second world war, the energy industries had been in private or municipal ownership. However the pressures to take these sectors into the public sector were considerable. The running of the coal industry was a focus for considerable industrial strife and trade unions and the Labour Party were committed to recasting the industry in public hands. Nationalisation was also seen as a way of bringing order to the energy utilities, especially electricity. A move in 1926 to create a publicly owned company to build and own a high voltage transmission grid and to coordinate power supply was only a partial success. The experience of wartime planning demonstrated both the potential of coordination as a solution to the different problems facing the energy sector and the limits of coordination without centralised ownership (Hannah, 1982).

Beyond the commitment to public ownership, however, most accounts agree on the rather unfocused nature of the nationalisation programme and its relationship to energy and economic policy. The underlying "Morrisonian" principle[2] of UK

---

1 For accounts of nationalisation, see Robson, 1962 and Kelf-Cohen, 1973. On the economic aspects of nationalisation see Foreham-Peck and Milward, 1994 and Foster, 1992.

2 Herbert Morrison was a Labour Minister in the 1930s and 40s and was responsible for nationalisation policy in both periods.

nationalisation was that public enterprises should operate at arm's length from government and operate in the public interest. The vagueness of this rationale in practice left industries largely unaccountable while permitting governments to pursue a range of objectives through public enterprises. Legislation establishing the industries was vague on actual commitments - beyond that of breaking even (one exception being the North of Scotland Hydro Electric Board which was established during the Second World War with the aim of fostering that region's development). Yet over the following years the energy industries were to be key players in economic reconstruction and retained a strategic aspect throughout the period of public ownership (Cairncross, 1985).

The involvement of energy industries in economic development - although of considerable importance in its own right - is not, however, normally regarded as the core concern of energy policy. Throughout this period energy policy was articulated and shaped in various governmental committees though it was not until the 1960s that an explicit statement of energy policy was made.[3] In the previous decades energy policy was mainly concerned with the tasks of postwar reconstruction and the identification and promotion of new resources and technologies.

The first formal White Paper on energy policy was published in 1965 and owed a lot to the then Labour government's reborn enthusiasm for planning. The document stressed the broader contribution to be made by the energy sector ("the strengthening of the economy and the balance of payments") as well as setting a number of objectives related to the maintenance of supply security, consumer choice and national competitiveness (Department of Fuel and Power, 1965). To pursue these objectives, the government established an Energy Advisory Committee (comprising the heads of the relevant nationalised industries, unions and government departments) in addition to a Coordinating Committee (consisting of the Ministry of Fuel and the energy industry chairmen). The 1967 White Paper broadly restated these objections but emphasised the development of the recently discovered resources of the North Sea (Department of Fuel and Power, 1967).

As noted, the White Papers came after a series of important initiatives in the energy sector over the 1950s and 1960s. Governments of both the left and right took key

---

3 The Simon Committee of 1946 and the Ridley Committee reported on fuel policy in 1946 and 1952 respectively.

decisions on nuclear industry development, coal industry rationalisation and North Sea oil and gas exploration thereby shaping the priorities of UK energy policy for the following decades. In many respects, therefore, the 1960s White Papers were as much rubber stamps on past events as blueprints for the future. While they laid out the rationale for policy and set a clear strategy on such issues as technology choice and depletion policy, in practice British energy policy was shaped by the interests of the government and the largely public sector energy industry.

Moreover, given the importance of public sector firms in carrying out policy, an important constraint upon energy policy was the evolving attempt by government to improve the performance of nationalised industries as a whole and those in the energy sector in particular. The general initiatives established the investment criteria for the public sector and placed greater emphasis on productivity and pricing behaviour. The more specific measures were designed to review the performance of enterprises and projects and to make improvements where necessary. In effect these measures constituted the government's attempt to regulate the public sector, a role which was carried out with mixed results.[4]

Over the following decade, the context to British energy policy changed dramatically though it was not until the mid 1970s that the government began a formal reappraisal of energy policy. The oil crises of the early 1970s revealed the vulnerability of the British economy to supply disruption and price escalation in the energy sector. A series of damaging strikes in the coal industry established the miners as an important veto-group in future energy policy and, by contributing to the downfall of the Conservative government in 1974, created a climate of hostility between that party and the coal industry which was to culminate in further conflict ten years later. Numerous debates on nuclear power revealed both the controversy of the technology in British society and the major differences which existed even amongst its supporters on how the industry should develop. More generally, however, there was a definite perception of energy as a valuable but diminishing natural resource which meant that energy policy decisions were taken within a scarcity culture. The conventional wisdom on energy policy was dominated by concerns over self sufficiency (though security of supply had always been an important aspect of

---

4 On the regulatory nature of the government-nationalised industry relationship see Rees, 1989 and Surrey, 1986.

policy), reinforcing the development of North Sea Oil, nuclear power and (at least for Labour) the future role of domestically produced coal.[5]

Throughout the 1970s a series of reviews and reports outlined the possible future paths for British energy policy, a debate which was supported by an influential Energy Commission in which energy industry views were prominent. The next formal statement of energy policy was the 1978 Green Paper based largely on an Energy Commission Paper which had recommended a considerable expansion of nuclear power and a major commitment to advanced nuclear technologies. While the Green Paper toned these down (notably with regard to advanced nuclear), the overall thrust of the new policy remained the same: the development of a range of indigenous energy sources backed by a substantial research commitment to medium to long term technologies.[6] This commitment was underpinned by a series of energy forecasts throughout the 1960s and 70s which indicated rapid growth in demand over the rest of the century (Baumgartner and Middtun, 1986).

For most of the period prior to 1979, therefore, the government conducted an official Energy Policy which was in many respects, highly effective, notably in the development of North Sea resources over the 1960s and 70s (both in offshore development and the conversion to natural gas). The mix of government ownership and clear objectives worked well in developing oil and gas reserves and in building a national network for natural gas. However from the 1970s on, government policy encountered more problems, most notably in the coal and nuclear sectors.[7]

In line with the character of economic policy at the time the main interests reflected in those policies were those of the government and the energy industries and their employees; consumer interests were (with the exception of very large industrial users) for the most part ignored despite the existence of active government-funded consumer organisations in both the gas and electricity sectors. Over this period, energy policy focused on two main issues: the development of North Sea resources (where the interests of the oil and gas industries prevailed) and the future of

---

5 See Cook and Surrey, 1977. On the adaptation of British energy policy in the 1970s see Chesshire, J et al, 1977.

6 Department of Energy, 1978. On the evolution of energy policy in the 1970s see Pearson, 1981.

7 See Ashworth, 1986 and Williams, 1980 on the problems of the coal and nuclear industries. For a critical view of nationalised industries, see Pryke, 1981.

electricity supply (where coal and nuclear power competed albeit with nuclear on the ascendant). In the latter case, the balance of power in energy policy making was evident from the relative weight given in both public financial support and utility strategy to coal and particularly nuclear technologies versus decentralised options such as renewables and Combined Heat and Power or energy efficiency (Rudig, 1986). The bias also reflected the dominance of central government over local interests: with nationalisation, local voices in energy decision-making were effectively silenced. Indeed, more generally, mechanisms of accountability for energy policy were not very well developed in the British system.

Nor was the balance of policy the only problem. Increasingly, the relative importance of formal energy policy began to diminish vis a vis other government policies. Informal pressures and the formal financial controls were probably the most important means to achieve a variety of economic goals extending well beyond the balance of payments and competitiveness concerns raised in the 1965 White Paper (including price control and the support of fuel and equipment supply industries for industrial and regional policy objectives). The value of these interventions may be questioned (particularly where short term political calculations dictated important decisions) but they cannot be ignored.[8]

The shifts in UK energy policy to some extent mirrored the changes in overall energy balances over the post war period (see Table 5.5). From being predominantly self sufficient and coal-based, the UK became steadily more dependent upon imported energy, mainly oil. As changing patterns of energy use developed (a move away from heavy industry and improved efficiencies with which energy was utilised), the UK economy became steadily less energy intensive. In particular it became less dependent upon coal notwithstanding attempts in the 1970s to encourage use of the fuel. The market for coal became increasingly focused on the electricity industry (where usage increased from 53m. tonnes in 1960 to 89m. tonnes in 1979). Here however its position was threatened by oil and nuclear from the 1960s onwards: by the end of the 1970s nuclear accounted for 13% of fuel input to power generation.

---

8 See for example the explicit policy goals set in the 1978 Green Paper. One particularly blatant example of how the energy sector was used for wider goals can be seen in the way in which the electricity industry was used to support the UK engineering industry. On the use of electricity industry to support other sectors see Surrey, Buckley and Robson, 1980.

Table 5.5: UK Energy Consumption: Primary and Final

mtoe	1960	1970	1980	1990	1994
Primary Energy Supply					
Coal	124.2	99.0	73.3	67.4	52.2
Oil	42.9	94.8	77.4	78.3	77.9
Gas	0.1	11.3	44.8	50.6	65.2
Electricity	0.9	7.45	10.2	17.7	23.2
Total	168.1	212.4	205.7	214.1	218.5
Import dependency (%)	27.6	47.9	6.4	2.2	-13.7
Final Consumption					
By Fuel:					
Coal	78.4	44.9	18.2	13.6	11.8
Oil	32.2	68.9	62.4	63.3	66.2
Gas	8.0	15.6	42.3	45.9	49.3
Electricity	8.5	16.5	19.3	23.6	24.8
Total	127.3	145.9	142.2	146.4	152.1
By Sector:					
Industry	53.8	62.3	48.3	37.8	37.5
Transport	22.2	28.2	35.4	48.6	50.6
Domestic	36.3	36.9	39.8	40.8	43.9
Other	14.9	18.6	18.7	19.2	20.1
Total	127.3	145.9	142.2	146.4	152.1

Source: UK Energy Digest, various

The other major change in UK energy balances, again reflecting energy policy decisions, was the development of offshore resources and, at least initially, the growing use of natural gas. By 1979 over 500 Twh of natural gas were consumed (80% of it produced in the UK), compared with 57 Gwh in 1969, and the transition from town gas to natural gas was more or less complete on the UK mainland. By the end of the 1970s North Sea oil production was beginning to reach significant levels, exceeding imports for the first time in 1979. Thanks to the development of North Sea resources, the UK had reduced its dependence on energy imports to the levels of the 1950s and self-sufficiency was in sight.

5.4 UK Energy Policy Since 1979

UK energy policy prior to 1979 - when the Conservatives returned to power - could be described as primarily concerned with the development of British energy resources with a view to meeting domestic energy needs but also with a view to fulfilling a number of broader economic objectives in developing high technology industries, supporting the engineering sector and promoting regional and social policies. In pursuing these goals, the publicly owned enterprises which dominated the energy sector played a critical role. In 1993 (after nearly 15 years in government) the government summarised its energy policy objectives as: to encourage competition amongst energy producers and to provide a regulatory framework to allow markets to work well; to commercialise energy markets in which the full costs of energy were borne by customers; to privatise the energy industries and support wider share ownership, to take account of the environmental impact of the energy sector and meet international commitments; to promote energy efficiency and safeguard health and safety (Department of Trade and Industry, 1993). While some aspects and aspirations are common to both periods, a change of emphasis - away from government planning towards the market - is all too apparent.

How did this transformation of policy come about? It would be a mistake to suppose that there was a sudden shift; just as "Thatcherism" did not constitute an overnight transformation of economic policy so energy policies changed gradually as new options emerged and old ones were discarded or downgraded. Indeed, one of the first energy policy decisions taken by the Thatcher government was a renewed commitment to nuclear power. At the end of 1979 the government announced a major investment programme in the Pressurised Water Reactor (PWR) (Times 19th December 1979). Such a commitment to a specific fuel option and a specific technology was characteristic of older policies and does not accord with the image of government disengaging from economic management. It has to be understood as primarily a continuation of existing policy though it also stemmed to some extent from Mrs Thatcher's own enthusiasm for the technology as a scientist and her desire to minimise the role of coal (which was seen as politically risky) in future power generation. In 1985 the government vetoed the proposed Sleipner project which would have linked the British and Norwegian gas fields, largely on the grounds of security of supply (Stern 1986). For much of the 1980s the government utilised controls on nationalised industries much more aggressively than its predecessors.

However, over the same period there were growing indications of a significant change in government attitudes as it gradually jettisoned various components of a traditional approach to energy policy. Although a separate Department of Energy was maintained throughout the 1980s, its main tasks were those of managing energy research and development and the offshore licensing system; traditional energy policy activities such as the preparation of forecasts were no longer pursued.[9] These omissions earned the government much criticism within the energy policy community in the UK and from bodies such as the International Energy Agency (Stern, 1987; IEA, various).

However, the most important shift was the increased emphasis on the market in the energy sector. The essence of this change was contained in a speech made in 1982 by the then Energy Minister, Nigel Lawson where he argued that the government's principal task was to "set a framework which will ensure that the market operates in the energy sector with a minimum of distortion and that energy is produced and consumed efficiently".[10] Such sentiments reflected the government's overall economic policy and its ideological commitment to private enterprise and competition (Gamble 1994). At the core of this shift in focus were the policies of privatisation and liberalisation. However, while the government claimed to be rolling back the state and promoting the role of the market, in practice policy initiatives were compromised partly by the power of incumbent interests in the energy sector but also by the government prioritising political expediency over economic efficiency. Although the change in emphasis was real enough and had dramatic impacts on the British energy sector, the nature of those changes was not always in line with the rhetoric of the period and appeared to reflect short-term political calculations..

---

9 To some extent, the shortcomings in the forecasting process in both government and energy industries can be regarded as discrediting the practice. See Stern, 1987 and MacKerron and Walker 1986.

10 Department of Energy, 1982. It is interesting to note that the shift in policy was marked by no more than a conference paper by the Energy Minister. Lawson himself revealed his hostility to traditional UK energy policy in his autobiography. See Lawson, 1992.

Privatisation, Competition and Regulation in the UK Energy Sector

Beginning in 1982 with the Oil and Gas (Enterprise) Act, the government embarked on a policy which was to lead to a radically restructured energy sector by the end of the decade. This first piece of legislation was designed to do two things, to divest British Gas of its oil interests, transferring them to the private sector, and to open up the British Gas pipeline system to allow other suppliers to compete with British Gas for final customers. However, the status of British Gas as a nationalised industry remained unchanged. The strength of British Gas as a public sector incumbent proved a major barrier to the success of the legislation. Partly this was due to the company's favourable position as a producer of gas: the cost of gas from its fields was on average approximately half that which any other competitor could supply. However, more importantly, British Gas was left to set the price for use of its transmission system. Given these problems it is perhaps unsurprising that there was such a lack of enthusiasm from potential competitors (Hammond, Helm and Thompson, 1985).

The following year, the government extended the policy of liberalisation to the electricity industry. The 1983 Energy Act abolished the electricity companies' statutory monopoly and obliged them to publish terms for the purchase and transmission of electricity. In principle anyone could build a power station and sell the power to the distribution companies or transfer the power across the national grid. However, this legislation met with only slightly more interest than the previous year's initiative. While the scanty use of the new provisions could be attributed to the overall competitiveness of the British electricity industry, as with the gas industry, the market power of the incumbents must also have deterred new entrants. The electricity companies were able to exploit their position as the setter of terms and conditions for purchasing electricity. This in turn was a result of the structure of the publicly owned industry with a vertically integrated production and transmission company - the Central Electricity Generating Board - dominating the system in England and Wales (Yarrow, 1989).

The government, therefore, was rather unsuccessful in the initial stages of its "market-based" energy policy. While legislation was introduced which opened up the possibility of limited competition in the traditionally monopolistic gas and electricity markets, the position of the incumbent publicly owned utilities neutered the policy from the start. The management within these companies was able to exploit their existing market power and their near monopoly of information on the industry's operations to render liberalisation ineffective. The power of the incumbent managements to limit the scope for competition in the energy sector was to be demonstrated again with the privatisation of British Gas.

By the mid 1980s the government had discovered the economic and political advantages of privatisation. After a shaky start - ironically regarding the sale of oil assets - the programme had achieved a number of successes, initially of public assets in the manufacturing sector. With the sale of British Telecom in 1984, however, the policy had entered a new phase, involving industries with a monopoly or near monopoly position. These sales had a number of advantages: they reduced government commitments to public funding of future investments; they raised revenue which the government could use to enhance its tax-cutting programme; they created a new constituency of support for the government amongst those who purchased the shares; and they boosted the financial services sector of the economy by establishing it as a centre of expertise in privatisation techniques. While many criticised the policy as merely transferring public monopolies into the private sector, the government and its supporters countered by arguing the virtues of private ownership, the possibilities of effective but light regulation and the potential for introducing competition into the sectors involved (Vickers and Yarrow, 1988).

Indeed, regulation and competition were intertwined in the new system of privatised utilities. The regulatory priorities were to be the promotion of competition where possible and the protection of customers where necessary. Mindful of the problems associated with regulation of privately owned utilities in the United States (such as poor incentives for efficiency and overgrown bureaucracies), the government sought to develop a new regulatory model. This involved the establishment of a specialised regulatory agency for the utility sector to be privatised which would oversee the behaviour of the firm or firms in that sector. At the core of the regulatory task was the setting of a price control which would have the effect of encouraging utilities to improve their efficiency and share the benefits of that improvement with customers. The so-called RPI-X formula allowed utilities to increase their prices by the rate of inflation minus an efficiency factor. In this way, customers would receive real reductions in prices and utilities would be obliged to improve their efficiency to remain profitable: if they failed to improve efficiency then they would not be able to recover the difference between costs and revenues from customers; if they succeeded they would be able to keep the additional savings as higher profits (Foster, 1992).

The system was seen as overcoming the problems of American-style rate of return regulation - which granted utilities a set level of return on their assets but which did little to control costs. It was moreover seen as a control for those customers which were not able to benefit from greater competition. As competition increased in the utility markets, it was argued, the need for such price regulation would wither away. Thus the other task of the regulator was to encourage the development of competition by ensuring fair access to utility networks and to prevent any anti-competitive conduct.

In most cases of privatisation, the use of regulation to encourage competition was the main mechanism for opening up markets. The alternative, of restructuring the utility before privatisation was generally rejected. This failure to dismantle the incumbent firms was the result of a number of factors, notably the uncertainty of new firm structures and of competition slowing down the pace of privatisation and reducing the returns from the sale. To the fore in making the case for retaining the existing shape of the industry were the managements of the utilities to be privatised (indeed senior management threatened to resign if the industry was reorganised). Thus when it came to be sold in 1986, British Gas was privatised as a monopoly utility with a regulatory agency (Ofgas) widely believed to be too weak to rein in the utility's conduct (MacKerron, McGowan and Surrey, 1990). It appeared that, far from being compatible and complementary strands of the government's new market-based energy policy, privatisation and liberalisation were effectively at odds with each other on many occasions Expediency in securing the best economic and political outcomes took precedence over the government's rhetoric of competition and the market.

Indeed, while the privatisation of British Gas was highly successful in financial and political terms, it proved controversial with many of the government's supporters and several of the country's largest gas consumers. The ensuing dispute between the company and the regulatory agency (which was to prove rather more robust than many had expected) was to last for the next eight years. Moreover, by that time the privatised utilities' performance was coming under increasing criticism, and their transfer from public ownership as monopolies was seen as one of the problems. In this context, it was hardly surprising that after the re-election of the Conservative government in June 1987, and with a commitment to privatise the electricity industry, the Government indicated that competition would be the key to the operation of the industry. Given that subsequent events hinge on the decisions surrounding that privatisation, it is worth recalling in some detail how the electricity industry came to be sold.

In the debate up to the publication of the White Paper on Electricity Privatisation there was considerable lobbying by the industry for the retention of the existing structure. They argued that a competitive structure which involved the disaggregation of the industry would be extremely costly. Yet, in the White Paper, the government rejected such arguments and resisted the pressure of the management to privatise in its existing form the company responsible for power production and transmission in England and Wales. Such an outright rejection of the incumbent company's lobbying was unprecedented in the eight years of the privatisation policy. The White Paper instead proposed the splitting of the Central Electricity Generating Board (CEGB) into three: a large power company owning 70% of production capacity including the system's nuclear power stations; a second production company owning 30%, and; a transmission company owned - as a separate venture - by the distribution companies (which were to be privatised in their existing form, following a rather more successful lobby against mergers or reorganisation) (Department of Energy, 1988).

By splitting up the task of power generation and by separating out the functions of production and transmission the government appeared to take heed of its critics and to have recognised the problems associated with previous liberalisations and privatisations. The establishment of transmission as a separate activity ostensibly levelled the playing field giving no one producer of power the undoubted advantage of being the carrier for all others. The disaggregation of the monopoly and competitive elements of the electricity industry - there was a similar split in the distribution business between delivery and sale of electricity - would, it was felt allow competition to prevail throughout the industry. In fact, the government sought to introduce competition at all levels of the industry: between suppliers of fuel and equipment to the industry; between the generating companies and between them and new independent producers of electricity in both the wholesale market for power (the electricity pool); between the generators and the distributors for customers with large energy requirements and; between distributors.

While the White Paper demonstrated that the government was willing to introduce a greater element of restructuring into the electricity industry than it had allowed in previous energy sector privatisations, there was no elaboration on how these attempts to increase competitive forces would work in practice. In a sense the White Paper was designed to answer one question ("what structure?") only to raise another ("how will it operate?"). As the government started to answer that second question there was a tension between political expediency (particularly regarding the need to complete the privatisation before the next election) and its own ideological commitments which was in the short term resolved in the former's favour: while competition was desirable, it generally took second place to the need to achieve a successful sale. For the most part, the solutions were stop-gaps and the failure to resolve them in a more lasting manner is at the root of the current crisis.

One should of course recall how novel the government's experiment in transforming the electricity industry was. The problem for the government in devising a competitive mechanism for the industry was that it was embarking on what was effectively an untried policy for the industry. Nowhere else in the world offered any experience in operating and planning an electricity system on competitive grounds. While competitive mechanisms did exist in some countries, they operated within primarily cooperative structures. The reason for the prevalence of cooperation and coordination over competition, according to most industry arguments, lay in the need to control the technical operation of the system, to minimise the operating costs of power plant and to prevent over capacity in the planning and construction of new capacity. The government effectively ignored such arguments as self-serving justifications for maintaining the old system, and claimed that new mechanisms could be devised to send the right signals to the industry. The measures adopted to meet those needs - primarily in the form of the pool pricing arrangements - have proved to be one source of the subsequent problems.

As the government began to develop its proposals - through the legislation and the operating licences which the various components of the electricity industry would require - and as the different parts of the industry sought to negotiate both the terms for the sale and purchase of electricity after privatisation and the details of the flotation of the companies themselves, the problems of introducing competition became ever clearer, most notably with regard to the risks which different parts of the industry would have to bear and the prospects for the coal industry. Precisely because the reform was potentially so radical, all those affected by the changes lobbied vigorously to minimise the scope of the changes or at least to slow their pace.

Faced with the problems of transforming the industry and the pressures of different interests, yet determined to privatise the industry before the next election, the government decided to impose a series of compromise agreements on the industry which limited the scope for competition in the period immediately after privatisation. The scope for competition amongst production companies and distribution companies was limited to a tranche of customers in excess of 1 MW until 1994, then a similar tranche of customers in excess of 100 kW until 1998 after which the whole system will be open to competition. Contracts for supply between distributors and generators were also signed for a limited period, dependent on a three year agreement between the coal industry and the generators.

Elsewhere in the electricity industry similar compromises emerged. The distribution companies gained a relatively light regulatory regime as plans to incorporate a system of "competition by comparison" into their price formulas were abandoned and the price controls themselves were set at modest levels (effectively imposing no real price reductions upon them). Protection also had to be secured for the nuclear industry which was revealed in the privatisation process to be much higher cost than the rest of the industry. Such were the problems that the government had to pull the nuclear industry from privatisation and devise a mechanism for ensuring its power was bought. The mechanism is twofold. First there is a non-fossil fuel obligation (NFFO) which compels the distributors to buy specified quantities of nuclear power, in practice amounting to whatever is the maximum current nuclear output. Second there is a so-called fossil fuel levy. This is a percentage charge on all sales of electricity, designed to compensate the distributors for the difference between the pool price and the price which Nuclear Electric charges (Vickers and Yarrow, 1991).

Over the following two years, the industry was largely privatised, though a 40% stake in each of the English generators remained in government hands until 1995.

The impact of privatisation was to be felt on other parts of the UK energy sector. The government's stakes in the oil sector were sold off over the course of the 1980s (including in 1987 a near-disastrous sale of British Petroleum in the wake of the stock market crash of that year). The re-election of the Conservatives in 1992 enabled them to fulfil their commitment to privatise the coal industry; the industry was sold to a private bidder in early 1995. In the same year it announced its plans to privatise Nuclear Electric in 1996/7.[11] Assuming these sales go forward the government will have moved from having a dominant stake in the British energy sector to having almost none.

Another aspect of the government's market based energy policy was the changing attitudes to the support of new energy technologies. In the late 1980s the government decided to reduce research and development funding to nuclear power and renewables. In the case of nuclear power, it initially limited its commitment to the electricity industry partly by transferring it to the CEGB where it was eventually scaled down as a result of privatisation and partly by cutting its own remaining budget (Cabinet Office, various). In the case of renewables the government announced a steady reduction in funds as technologies either proved themselves and were taken up by the private sector or proved too long term and were dropped. Although indirect support was offered through the Non Fossil Fuel Obligation, the thrust of government policy remains that of "market enablement" with a view to abandoning its involvement early in the next century (Department of Energy, 1988; Department of Trade and Industry 1994). The emphasis on markets was also reflected in the government's claims for its approach to assessing and tackling environmental problems though, as will be seen, the political risks of applying market-mechanisms in dealing with these problems effectively outweighed ideological considerations.

Beyond the Market - the Persistence of Old Agendas and the Emergence of New Ones

Of course, privatisation and competition did not constitute the whole of energy policy under the Conservatives: as noted, the treatment of nuclear power in the early 1980s and the Sleipner decision of 1985 were just two areas where a mix of

---

11 Ironically, the company is to be re-named "British Energy". The older nuclear power stations will be left in public ownership.

traditional concerns and political preferences shaped government responses. Policy towards energy efficiency was another area where government policy remained relatively consistent for much of the period. Through R&D funding on both the supply and demand sides and a series of information programmes (often involving local government as a target and or partner for initiatives), the government maintained a commitment to improving UK energy efficiency. However, the extent of that commitment was subject to criticism throughout much of the period. Critics pointed to the privileged position of investments in new energy projects vis a vis energy efficiency projects and the government's failure to address this market failure (Chesshire 1986).

Hopes for a more robust policy on energy efficiency were given a boost by two developments: the establishment of regulatory agencies to monitor the gas and electricity industries and the growing importance of the environment in energy policy calculations. The legislation privatising the energy utilities included commitments to energy efficiency. Although much vaguer than the proponents of energy efficiency would have wanted they provided a basis for lobbying the regulatory agencies once in operation. While the regulators took very different approaches to energy efficiency (see below) they did back the establishment of an Energy Savings Trust (a move also supported by the privatised utilities and the government). The Trust was to coordinate energy efficiency projects which would be funded out of a levy on customers. However a combination of factors put paid to an extensive programme and the Trust appears likely to operate on a more modest scale.[12]

Much of the impetus for the Trust and for the continued visibility of the energy efficiency issue as a whole was due to the impact of growing concerns over the environmental consequences of energy production and use. Indeed it could be argued that without the emergence of the environment as a constraint upon energy policy-making, the changes in energy markets witnessed in the mid 1980s would have rendered redundant much of the rationale for not only energy efficiency but also other energy options such as renewable energy and nuclear power..

To some extent the environment had always been an issue in energy policy, but before the 1980s it was discussed as much in terms of the depletion of natural

---

12 Financial Times 3rd February 1995

resources as in terms of the consequences of their use, with the latter mainly conceived of from the point of view of catastrophic disasters (oil spills, nuclear accidents , etc.). What happened over the 1980s was a recognition that day to day energy use could damage the environment, first in regional terms (as with questions of acid rain) and later in global terms. The translation of that recognition into an energy policy considerations took longer in the UK than other European states. For many years the government had been extremely defensive on environmental issues (which were themselves dealt with by lengthy Royal Commissions or public inquiries, mechanisms which effectively neutered their political impact).

However, growing awareness of the environmental consequences of producing and consuming energy (not least on emissions, see Table 5.6) and the rise of green politics in Britain, influenced the tone of policy. The Conservative government reinvented itself as concerned for the environment, publishing a White Paper on the Environment (Department of the Environment 1990). The White Paper set out to address a number of environmental issues, not just those related to the energy sector but the sector figured prominently. In it the Government discussed a number of options for the future, including the encouragement of energy efficiency and renewable energy as well as further scrutiny of the development of the transport sector. Nonetheless, major problems related to the environment remained to be resolved. In particular the attempt to reconcile the government's commitment to the market with the environment by devising economic mechanisms which would internalise the external costs of energy supply and consumption, proved to be difficult proved easier said than done. While a policy of environmental taxes and "polluter pays" charges might have been consistent with the government's ideology, the practical difficulties and political dangers of implementation were too great.

Table 5.6: **UK Emission Levels**

m.tonnes	1970	1980	1990	1993
Carbon dioxide	182	164	158	152
Sulphur dioxide	6.5	4.9	3.8	3.2
Oxides of nitrogen	2.3	2.4	2.9	2.4

Source: UK Digest of Environmental Protection and Water Statistics

The rise of the environment as an energy policy issue drew attention to the condition of British energy balances (incidentally requiring the government to engage in a process of detailed energy forecasts for the first time in many years). It was clear that what fuels were being produced and utilised in the UK would affect how well it could meet its international commitments. Those commitments were also beginning to impact upon what energy options were being chosen (the increased use of gas in power generation will make a significant contribution to efforts to reduce emissions).

In any case over the period of Conservative energy policy, the energy landscape had been transformed. Coming to power in the wake of the second oil shock the government found itself in the position of being self-sufficient for energy (though the economic effects of this situation have been much debated in the UK). From that point of view it could be argued that the government was able to afford the luxury of a radical reorientation of energy policy. Beyond this, moreover, the subsequent easing of energy markets (and the relatively modest disruption which accompanied the Gulf Conflict) arguably reinforced the implication underlying government policy - that energy was only a commodity and not subject to special treatment.

In energy balance terms the changes were as radical in this period as they had been in the preceding post war period. The collapse of oil prices severely damaged coal's position in the UK energy economy increasing its dependency on the electricity market. Yet at the same time the electricity market was to change radically (see below) and by the end of the period, coal accounted for only 50% of fuel input (a significant proportion of it imported) losing out to nuclear power and most recently natural gas. The shift was ironic since the coal industry had undergone a massive increase in efficiency over the years since the miners' strike. Although employment was down, output was broadly stable reflecting a radical improvement in productivity. These transformations caused many to argue that the UK would rapidly return to dependency on energy imports but while there has been a slight shift to dependency in recent years as a result of major disruptions to oil production following the Brent Spar disaster and the growth of coal and electricity imports) it remains at a low level (see Table 5.5).

As noted earlier, the government failed to provide any official forecasts of UK energy balances for much of the 1980s and it was only the need to meet environmental obligations - primarily with regard to controlling emissions of carbon dioxide that the government published a new set of projections (initially in 1992 but revised in 1995 - see Table 5.7).[13] The forecasts comprised a set of six projections associated with various scenarios of economic growth and energy prices, though they also incorporated a number of policy assumptions (such as the effect of the government's energy efficiency initiatives and the nature of investment in the electricity supply industry).

Table 5.7:   UK Energy - Projections to 2020

mtoe/m.tonnes	Primary Energy Requirements			Final Energy Consumption			Carbon Dioxide Emissions		
	1990	2000	2020	1990	2000	2020	1990	2000	2020
Low growth low price	221	231	262	147	162	195	158	147	171
Low growth high price	221	226	253	147	158	187	158	144	173
Central growth low price	221	237	283	147	166	210	158	150	184
Central growth high price	221	232	273	147	161	203	158	148	188
High growth low price	221	240	298	147	169	215	158	152	193
High growth high price	221	235	289	147	164	215	158	151	197

Source: Department of Trade and Industry, Energy Paper 65

5.5   British Energy Policy in Crisis?

By the early 1990s, the priorities and processes of British energy policy were very different from what they had been before 1979. While it took time to emerge, and while it inevitably was shaped by wider political goals, the policy did mark a departure both for the government and the sector. As such it served as a useful corrective to some of the previous shortcomings in previous energy policies, introducing greater transparency into some aspects of the industries' operations (though not necessarily in terms of information available to the public), giving more weight to consumer interests and creating a formal regulatory system. Equally, however, the new policy raised a number of problems. The pursuit of a market

---

13 See Department of Trade and Industry Energy, 1995. The environmental rationale is spelt out here though it also notes the importance of forecasts for assessing the possible role of nuclear power.

based energy policy was proving to be more difficult to implement than expected and privatisation of the electricity industry had only been possible as a result of a number of expedient moves (on nuclear power on coal and on regulation). While difficulties persisted, the government maintained its commitment to competition, privatisation and non-intervention, as symbolised by the closure of the Department of Energy after the election in 1992. Nor was there any reason to suppose that pursuing the policy further would create political difficulties. However over the next few years the government encountered a number of highly controversial and, in some cases, politically damaging crises.

Gas and Electricity Liberalisation - Competition and its Consequences

To understand the events of that period, one has to look more closely at both the gas and electricity industries post privatisation and particularly the respective roles of the regulators in those industries. To a large extent the roots of the problems which arose in the early 1990s can be attributed to the structures established at the time of privatisation and the efforts of the regulators to adjust them. A common theme in both cases was the attempt to promote competition.

Although gas was privatised intact, there was almost immediately pressure from consumers, potential competitors and the regulator to limit the industry's power. Almost from the outset, the Director General of Gas Supply, James MacKinnon was effectively allaying widespread fears that he would be a weak regulator. Buttressed by the results of an early reference to the Monopolies and Mergers Commission (MMC) which recommended that British Gas should abandon a number of anti-competitive practices, Ofgas began to push for greater competition in the gas industry. A central element of the reforms was the requirement to publish a schedule of the prices at which British Gas was prepared to supply firm and interruptible gas to contract customers and to open up to competitive supply (MacKerron, McGowan and Surrey 1990). The moves did not resolve the issue of competition in gas markets, however. Instead conflict between Ofgas and British Gas continued, leading to further interventions by the British competition authorities.

However, while the dispute over competition continued for some years (Parker and Surrey 1994), they undoubtedly rendered gas a much more attractive fuel for power generation. It was already seen to have a number of advantages (environmentally friendly and low capital cost). For both the existing power producers and potential new entrants, therefore, gas was seen as the most attractive option for new power plant. There was moreover considerable interest in new power stations because the structure of the electricity industry was seen as in many ways scarcely more competitive, at least vis a vis generation. The two privatised generation companies were regarded as exercising a duopoly which could only be countered by investment by distribution companies in independent power ventures. Such moves towards competition were of course part of the government's intention in the 1988 White Paper.

These moves were encouraged by the regulatory structure in the electricity industry and particularly by the regulator, Stephen Littlechild (also the architect of the UK regulatory system for privatised utilities). Coinciding with the effective deregulation of gas supply it became possible for power producers to purchase gas. The scale of commitment to using gas for new projects was, however, much greater than was expected. Within two years of privatisation and with new contracts looming it became clear that a massive investment in new capacity was planned by a mixture of existing generators, non utility producers and consortia involving distribution companies. The participation of this third group was particularly important as many of the new power stations were supplied gas on the basis of take or pay contracts which necessitated similarly strong commitments to purchase the electricity produced.

The emergence of this surplus of gas based power capacity and the apparent ring fencing of nuclear left only one possible target for cut-back: coal fired generation. Given the determination of the major generators to increase their purchases of cheaper imports of coal, it appeared that UK deep mined coal would be the main target. The negotiations for a new contract between British Coal and the generators indicated that much lower amounts of coal would be contracted for, leading to substantial closures in the UK industry. The rationalisation programme was announced in October 1992, provoking anger not only in the coal industry and the political opposition but also within the government and in the country at large. For some time it appeared that, so great was the political uproar from supporters as well as opponents of the government, its survival was at risk.

The political crisis engendered by the announcement was surprising enough. What was even more surprising was that the frustration did not simply focus on the effects of the closures on the communities involved (though this was undoubtedly an important dimension) or on the fact that the closures were prompted by the nature of the new arrangements in the electricity industry. Almost as significant in the public's perception was the sense that policy was adrift, that things would have been different if a coherent energy policy was in place. The government's attempt to distance itself from the consequences of its policy backfired. The director of the Office for Electricity Regulation made it very clear that his obligations towards the electricity industry did not extend to supporting the coal industry and that any such action had to be taken by the government (Littlechild 1993).

As it turned out the government was able to avoid defeat by announcing a major review of the British coal industry and of energy policy, a move which was effective in quelling the momentum of events. The White Paper which appeared early in 1992 made much of the difficulties of reversing the policy - largely on the grounds of international commitments - and offered a limited package of subsidies to support the industry, subsidies which were not for the most part utilised. Instead the rationalisation of the coal industry which had caused so much controversy some months earlier took place and the government maintained its policy, explicitly putting competition and privatisation at the centre piece of its policy and downgrading and even abandoning other obligations.[14]

Although it survived the crisis, the government had been obliged to confront the political consequences of its new policy. Competition and regulation could not be relied on to address all energy policy problems. Indeed, whether or not the government or other interested parties might have had that regulation would act as a surrogate for energy policy were to be disabused over the following years. In the field of energy efficiency, for example, both gas and electricity regulators indicated that they were not prepared to take responsibility for this policy: the electricity regulator made clear his misgivings from an early stage while the successor to Sir James McKinnon, Clare Spottiswoode, announced that she was not prepared to "tax" gas consumers in pursuit of energy efficiency (Financial Times 28th April 1994).

---

14 Department of Trade and Industry, 1993. This has been restated in the government's annual review of energy, Department of Trade and Industry, 1994.

Energy Policy and Regulation

However, beyond the question of whether the regulatory agencies should act as a surrogate energy policy-makers lies a more fundamental question over the nature of regulation itself. Almost from the moment they were established those responsible for regulation of the energy industries have been criticised. The substance of the criticisms, however, varies quite widely according to who is making them (producers or consumers) and when they were made (in the early years there were worries that the regulators were too weak, more recently that they were too powerful). To some extent they focus on particular aspects of the regulatory process (the nature of price control formulas, the emphasis given to competition, etc). Here, however, we note some of the broader problems which have been identified with this system of energy regulation.

When the privatisation of British Gas took place, the predominant concern was whether the regulatory regime, particularly the Office of Gas Supply, would be effective and strong enough to scrutinise the firm which dominated the gas market. Critics pointed to a lack of resources (approximately twenty people worked for the agency), the dominance of government officials within the agency (who it was thought would not be tough enough with British Gas) and the lack of relevant experience of the director of the agency (who had come from a private sector background and had no grounding in energy matters). It was also clear that British Gas itself did not take the regulator very seriously in the first instance, though subsequent events were to prove that this was an underestimation.

The question of the effectiveness of regulation also arose in the electricity sector. The regulator has been criticised for being too weak with the electricity utilities particularly when price control reviews have been undertaken. The regulator's decision to overturn his review of distribution price controls in early 1995 (following the publication of financial details in the wake of an attempted take-over in the sector) illustrated the problems of obtaining information from the regulated industries. Such an imbalance in information is regarded as a classic symptom of regulatory capture, where the regulator effectively protects the industry he/she is supposed to be regulating (Stigler, 1973). The danger of such capture is arguably reinforced by the nature of regulatory obligations which require the regulator to safeguard the interests of the industry as well as of the consumer. Many believe that the resulting balancing act has been to the detriment of the consumer and call for greater resources for the regulatory agencies and a greater emphasis on consumer interests .

From an energy industry point of view, however, there is a widespread concern about the strength rather than the weakness of the regulator. For many the regulatory agencies - and particularly the regulators - have too much power and discretion in carrying out their tasks. This imposes considerable uncertainty upon the industry as the regulator's actions become unpredictable. This was certainly a criticism of the Office of Gas Supply as it moved towards encouraging greater competition in the gas market. Relations between the regulator and the regulated became so bad that British Gas requested that the British competition authorities review the case. The concentration of power in the hands of the regulator has also led to allegations of a "cult of personality" whereby the preferences and prejudices of that regulator appear to affect the conduct of regulation more than the written rules. Various solutions - such as extensive judicial review procedures, replacing regulators with US-style panels of regulators, making regulators accountable to Parliament - have been put forward to limit regulatory arbitrariness.[15]

Whether they hold the regulator to be too powerful or too weak, such criticisms reflect a regulatory dilemma - what should be the balance between effectiveness and accountability? It could be argued that the requirements for effectiveness include endowing the regulator with substantial powers and the discretion to implement them as he/she sees fit. Such actions may however undermine the task of "regulating the regulator": how to ensure that the regulator acts responsibly. Indeed, beyond the implications of such a system for energy producers and consumers and even for energy policy, such a division of powers raises questions about democratic procedures and the extent of political control upon the regulatory process.

The British model of regulation has changed the way in which public interests are represented in the organisation and operation of energy industries and markets. By creating formal procedures and institutions enjoying significant autonomy, government has put itself at one remove from the energy sector. It would of course

---

15 On the problems of utility regulation and some proposals for reform see Veljanowski, 1993; National Consumer Council, 1993 and Corry, 1995.

be wrong to say that some of the problems and issues which arise in the new system are completely new in themselves. The issue of regulator-regulated relationships and of capture arose in the dealings between government ministries and nationalised industries prior to privatisation and persist in countries where regulation is not such a clear cut process (Foster, 1992; Rees 1989).

5.6   British and European Energy Policy - Leader or Laggard?

The European Union acts as a constraint upon British energy policy-making as it does for all member states.   However, the British attitude towards European initiatives in the energy sector is rather different from other parts of the Union. Partly this is a result of the UK's endowment of energy resources but it is also a result of the particular policies adopted in the UK itself over the last fifteen or so years.

As the only member of the European Union to be self-sufficient in energy resources, the UK has generally been unwilling to surrender sovereignty over those resources to the Brussels.   This was particularly the case in the 1970s when - with North Sea oil and gas resources coming on stream in the middle of a severe energy crisis - the UK resisted any attempt by the Commission to develop a common energy strategy. The resistance of any encroachment on the management of resources persisted even into the 1980s when it might have been argued that the UK government had itself abdicated responsibility for these matters.   Generally therefore the UK's position as a producer of energy resources has left it hostile to attempts to develop a European energy policy.

As we have seen, however, UK exceptionalism extends beyond its position in energy markets to its approach to energy policy.   With the emphasis on competition and privatisation, the scaling down of public funding for energy research and the neglect of energy planning and forecasting in the 1980s, the "private good - public bad" perception of energy policy did not endear the British government to traditional energy policy agendas and techniques or to some of the newer activities in the energy sector such as the environment. The UK government has for example been suspicious of the European Commission's initiative on Trans European Networks because of its associations with planning and economic pump-priming (House of Lords, 1993)

The area which has presented the greatest difficulties between the UK on the one hand and the Commission and many of its fellow member states on the other has been the environment. Although the British government adopted green rhetoric in the late 1980s this followed a long period of strained relations with some states on the UK's environmental record, reaching its nadir in the 1980s in a dispute between the government and Scandinavian states over emissions which caused acid rain (Boehmer-Christiansen and Skea 1990; Weale 1992). More recently, it has been the UK's attitude to greenhouse gas emissions which have caused problems. The government has been the main opponent of proposals for a carbon tax on energy, effectively blocking progress in this area (Financial Times, 27th April 1993).

However, while it is undoubtedly the case that the UK did not welcome European initiatives on energy policy, it was less hostile to other aspects of European policy towards the energy sector. The principal example of a policy where the UK has been not only supportive of Commission proposals but has acted as an exemplar for the policy is that of energy market liberalisation. The Commissions attempts at liberalising gas and electricity systems have been marked by considerable debate and opposition from many member states. It is fair to say that the UK has been useful to the Commission both as an exemplar, proving that liberalisation could be done [16], and as a supporter of Commission actions within the Council. The UK was also key to promoting the idea of a European Energy Charter as a market-based initiative. Early in the debate the UK advocated an emphasis on the role of the private sector and markets to encourage energy industry development.

Given its role in this field it is perhaps ironic that the privatisation of the British electricity industry was to provide the Commission with an opportunity to intervene in national energy policy-making. However, tackling an industry in the throes of change is from the Commission's point of view much easier than addressing the much greater forces which are not changing. While basically rubber stamping the planned contracts between the different parts of the industry after privatisation, the Commission required major changes in the planned support of the nuclear industry, imposing the limited subsidy and protection regime that is currently in force. Such moves have not endeared the Commission to the UK despite similar goals. As a

---

[16] In its election manifesto in 1992, the government made much of how other European countries were copying its policies. See also Department of Trade and Industry, 1994.

result the British remain firm supporters of subsidiarity even where it limits the scope for liberalisation.

5.7   Conclusion

British energy policy has undergone radical changes in the last fifteen years. From being a policy dominated by public enterprises and government funding aiming to coordinate long term strategy, energy policy is now characterised by private firms competing in relatively open markets and by the absence of a well articulated government role. Government continues to intervene but its activities are much less pervasive than in the past and to the extent that public authorities are active in the sector most of the time it is the regulatory agencies which are to the fore, operating on an agenda far narrower than what used to be considered as energy policy's realm. Ignoring the short run political calculations of particular acts like gas and electricity privatisation on the one hand and the ideological trappings of the shift on the other, the government's redefinition of energy policy has a wider benefit. Energy Policy was largely what a cartel of producers and large consumers wanted coinciding with the immediate political needs of governments and bureaucrats. The strategic dimensions of policy can with hindsight be viewed as at best misconceived and at worst promoting rather narrow interests: hyperopia can be just as bad as myopia. In that sense, the changes in policy were a corrective to the shortcomings and excesses of before.

That is not to argue that the new policy has been an unqualified success, however. Indeed how one views the effectiveness of energy policy in the last few years depends very much on where one sits. Investors and senior managements in the energy industries as well as their advisors have done very well from the process of privatisation and liberalisation. The market value of the electricity industry alone has almost tripled in the few years since privatisation, a development which has very little to do with performance since privatisation and very much to do with the way in which the industry was sold. In effect there has been a major transfer of resources from the previous owners of the British energy sector (the British taxpayer) to the new owners. The industries themselves have undergone massive rationalisation as the industries have limited their commitments to research, local suppliers, etc. This trend is motivated partly by efficiency constraints in the regulatory process and the pressure of competition (McGowan 1995). Yet the consumer - the ostensible beneficiary of greater competition - has seen only modest price changes.[17]   The regulatory process itself has had a mixed record, with early effectiveness in the gas industry and misjudgements in the electricity sector prompting calls for reform from inside and outside the industry. The narrower focus of policy has to some extent been limited by the growth of environmental concerns (where old energy policy agendas have to some extent found new homes) but even here many commitments are being met by fortuitous changes in energy markets rather than by a coherent strategy

Yet despite its problems, there are unlikely to be major changes in British energy policy in the near future. With energy balances remaining strong and energy prices remaining low, there are few pressures for taking a strategic view of energy. The regulation of the energy utilities is likely to present problems for regulators and governments alike (particularly in once competition is extended to all customers) but the risks of radical reform are likely to work against major changes of focus (though there may be institutional innovations such as a merging of Ofgas and Offer). Certainly the political opposition in the UK has no plans to return the sector to public ownership or to tinker radically with the sector. [18] Indeed, by analogy with the post war years - when nationalisation was accepted by all major political actors - the market-based energy policy is part of the neoliberal consensus that currently characterises the UK. As such it remains at odds with the way energy policy is pursued elsewhere in the European Union. While the British government's emphasis on liberalisation is in line with European Commission proposals for some energy sectors, the overall approach to energy policy remains much more limited than that envisaged by Brussels. The UK is therefore likely to remain on the fringes of European energy policy in more senses than one.

---

17 While gas prices are lower than before privatisation this is largely a result of the indexation the price control to world oil prices. In the electricity sector prices were increased substantially in the run-up to privatisation, a move which casts doubt on any claims to price reductions since privatisation. Studies of electricity prices moreover suggest that prices would have been lower had the industry remained in public hands. See National Consumers Council, 1994.

18 The Labour Party has abandoned any plans to take any part of the energy sector back into public ownership and recent statements on economic policy suggest a recognition of the virtues of competition even in the utility sector. See McGowan, 1995b. On energy policy per se, recent thinking suggests no radical changes are likely though there may be some innovations particularly with regard to energy conservation. For one contribution to the debate see Jones 1994.

References

Ashworth, W The History of the British Coal Industry, Vol 5, Clarendon, 1986.

Baumgartner, T and Middtun, A (eds) The Politics of Energy Forecasting, Clarendon 1986.

Boehmer-Christiansen, S and Skea, J Acid Politics, Bellhaven, 1990.

Cabinet Office Annual Review of Government Funded R & D, HMSO, various..

Cairncross, A Years of Recovery, Methuen 1985.

Chesshire, J "An Energy Efficiency Future - A Strategy for the UK" Energy Policy Vol 14 No 5 1986.

Chesshire, J et al "Energy Policy in Britain" in Lindberg, L (ed) The Energy Syndrome, Lexington Books, 1977.

Cook, L and Surrey, A J Energy Policy, Robertson, 1967.

Corry, D(ed) Profiting from the Utilities, IPPR, 1995.

Department of Energy, Privatising Electricity, Cmnd 322, HMSO, 1988.

Department of Energy, "Energy Policy - A Consultative Document" Cmnd 7101, HMSO, 1978.

Department of Energy "Renewable Energy in the UK: the Way Forward" Energy Paper 55, HMSO, 1988.

Department of Energy "Speech on Energy Policy" Energy Paper 51, HMSO 1982.

Department of Fuel and Power, Fuel Policy, Cmnd 3438, HMSO, 1967.

Department of Fuel and Power, Fuel Policy, Cmnd 1798, HMSO, 1965.

Department of the Environment, This Common Inheritance, Cmnd 1200, HMSO, 1990.

Department of Trade and Industry, Energy Report, HMSO, 1994.

Department of Trade and Industry, Energy Report, HMSO, 1995.

Department of Trade and Industry, British Energy Policy and the Market for Coal, Cmnd 2235, HMSO, 1993.

Department of trade and Industry, "New and Renewable Energy: Future Prospects in the UK" Energy Paper 62, HMSO 1994.

Department of Trade and Industry Energy Projections for the UK, Energy Paper 65, HMSO, 1995.

Foreham-Peck, J and Milward, R Public and Private Ownership of British Industry, Clarendon 1994.

Foster, C Privatisation, Public Ownership and the Regulation of Natural Monopoly, Blackwells 1992.

Gamble, A The Free Economy and the Strong State, Macmillan, 1994.

Hammond, E Helm, D and Thompson, D "British Gas: Options for Privatisation" Fiscal Studies, Vol 6 No 4 1985.

Hannah, L Engineers Managers and Politicians, Macmillan, 1982.

Hannah, L Electricity Before Nationalisation, Macmillan 1979.

House of Lords Select Committee on the European Communities, Structure of the Single Market for Energy, House of Lords Paper 56, London: HMSO, 1993.

International Energy Agency (IEA), Energy Policies and Programmes in IEA Countries, IEA/OECD, various.

Jones, D Energy Policy Now, Institute for Public Policy Research, 1994.

Kelf-Cohen, R British Nationalisation 1945-73, Macmillan 1973.

Lawson, N The View from Number 11, Bantam, 1992.

Littlechild, S "The Supply Price Control - Issues for Review" in Gilland, T (ed) Regulatory Policy and the Energy Sector, CRI, 1993.

MacKerron, G McGowan, F and Surrey, A Regulating the Privatised UK Gas Industry, SPRU, 1990.

MacKerron, G and Walker, W "Energy Forecasting: has it a Future?" in Gretton, J and Harrison, A (ed) Energy UK 1986, Policy Journals 1986.

McGowan, F and MacKerron, G "Contractualisation in the UK Electricity Industry" in Harrison, A (ed) From Hierarchy to Contract, Policy Journals, 1993.McGowan, F "Utility reform in the UK - the role of regulation and the impact on public service "Annals of Co-operative and Public Economics, 1995a.

McGowan, F "Labour's New Competition Policy: Market Forces and Public Interests" Renewal, 1995b.

National Consumer Council, Paying the Price: a Consumer View of Water Gas Electricity and Telephone Regulation, HMSO, London, 1993.

Parker, M and Surrey, J UK Gas Policy: Regulated Monopoly or Managed Competition, SPRU, 1994.

Pearson, L The Organisation of the Energy Industry, Macmillan 1981.

Political and Economic Planning A Fuel Policy for Britain, PEP 1966.

Pryke, R Nationalised Industries, Robertson 1981.

Rees, R "Modelling Public Enterprise Performance" in Helm, D Kay, J and Thompson, D The Market for Energy, Clarendon, 1989.

Robson, W Nationalised Industry and Public Ownership, Allen and Unwin, 1962.

Rudig, W "Energy Conservation and Electricity Utilities: a Comparative Analysis of Organisational Obstacles to CHP/DH" Energy Policy Vol 14 No2 1986.

Stern, J UK Energy Issues 1987-92, Joint Energy Programme Occasional Paper, 1987.

Stern, J Natural Gas in the UK: Options to 2000, Gower, 1986.

Stigler, G "The Economic Theory Of Regulation" Bell Journal of Economics, Vol 2 1973.

Surrey, A J Buckley C M and Robson, M "Heavy Electrical Plant" in Pavitt, K Technical Innovation and British Economic Performance, Macmillan 1980.

Surrey, A J "Government and the nationalised Energy Industries" in Gretton, J and Harrison, A (eds) Energy UK 1986, Policy Journals, 1986.

Veljanowski, C The Future of Industry Regulation, European Policy Forum, 1993.

Vickers, J and Yarrow, G "The British Electricity Experiment" Economic Policy no12 1991.

Vickers, J and Yarrow, G Privatisation - an Economic Analysis, MIT 1988.

Weale, A The Politics of Pollution, Manchester UP, 1992.

Williams, R The Nuclear Power Decisions - British Policies 1953-78, Croom Helm, 1980.

Yarrow, G "Regulatory Issues in the Electricity Supply Industry" in Helm, Kay and Thompson, 1989.

6. The Future of EU Energy Policy

Dominique Finon and John Surrey

IEPE   Science Policy Research Unit
University of Grenoble University of Sussex
Grenoble   Falmer, East Sussex
France   United Kingdom

6.1 Introduction

The Commission of the European Union (EU) is conducting a review of energy policy which will result in a White Paper for the Intergovernmental Conference in 1996. This adds to the timeliness of an independent assessment of the achievements and future scope of EU energy policy, which is the aim of this paper. It is of course relatively easy to note the objectives of EU energy policy. Briefly stated, they are to abolish the remaining barriers to trade and competition and 'complete the internal market'; to reduce emissions of 'acid rain' and 'greenhouse gases'; and to improve longer-term supply security.

Since the early-1970s the European Commission made a number of attempts, which proved largely abortive, to establish a common energy policy. But in the 1980s, in the wake of a revival of the EU (via the Single European Act, the Single Market programme and the more assertive application of European Competition Policy) the Commission embarked on a series of initiatives to liberalise the energy sector.

Over the same period, environmental protection also influenced policies towards the energy sector along with more traditional concerns over supply security. After four years of negotiations the member states agreed to the Large Combustion Plant Directive in late-1988 to regulate SO_2 and NO_x emissions from power stations and large industrial boilers. With this, the European Community acquired authority for policy to limit trans-boundary air pollution. The development of European environmental policy coincided with growing concerns over global warming. Several member states committed themselves to strong measures to reduce CO_2 emissions and wanted other member states to follow suit, and - when it seemed desirable - to have a common stance in international negotiations leading up to the

1992 Global Climate Change Convention in Rio.   The member states sought to establish common ground amongst themselves on measures to promote more efficient use of energy, the development of renewable energy sources, and to change the fuel supply mix in order to reduce CO_2 emissions.

At the same time as the Commission tried to take account of global warming, the old concern of supply security suddenly reappeared on the agenda after the dissolution of the Soviet Union, the collapse of communism in Eastern Europe, Iraq's invasion of Kuwait, disputes between the countries traversed by the pipelines bringing oil and gas to the West from the FSU (former Soviet Union), and the religious and political disturbances in Algeria.   These events underlined once more Europe's vulnerability to disruptions in international energy supplies, but the continuing abundance of oil and gas supplies (despite the long UN imposed ban on Iraq oil exports) meant that this concern was short-lived.

Two further factors have influenced the scope of European energy policy.   One is the major differences between Member States in energy requirements, energy import dependence, and the social and political importance of their fuel industries.   These have tended to limit European energy policy to common principles and objectives, regular talks with major foreign oil and gas producers, and support for energy R&D (including a very costly fusion programme).   In some cases the principles enunciated by the Commission have had little real substance (such as in the 1970s with the introduction of European rules to limit the use of oil and gas in power generation when high oil and gas prices were already a considerable deterrent).

As a rule, when important national interests have been at stake, member states have been reluctant to abandon national policies and practices which protect their own fuel industries, while other member states which are heavily dependent on imported energy will not collaborate in any scheme which would make them buy fuel at above world market prices.   For the most part, therefore, the locus of decision making on energy policy remained with the member countries; the periodic attempts by the Commission to formulate statements on energy policy objectives were largely summations of national forecasts and goals.   It remains to be seen whether or not this dominance of national interests persists.

The other influential factor is the uncertain future constitutional development of the EU despite the fact that the Maastricht Treaty appeared at the time to mark decisive progress towards economic and monetary union en route to a European federal state. The subsequent difficulties of ratification in several member states revealed deep splits in public opinion between the objectives of unification among existing EU member countries and a looser confederal association of many more nations, including the East European countries. Given this major uncertainty at the heart of the EU, it would not be surprising if the European Commission chose not to jeopardise the widest possible support on the central issue of integration and does not push too hard on sectoral policies which are opposed by key member states. In addition, member states are reluctant to 'cede sovereignty' (as they see it) in order to support EU energy policy objectives while the fundamental constitutional uncertainty remains. This reluctance is important if we bear in mind that proposals by the European Commission require the approval of the Council of Ministers (consisting of representatives of member state governments). Despite increased use of majority voting to restrict the use of the veto, there is still considerable scope for the pursuit of national interest, both via the Council of Ministers and in other ways. This is reflected in the many derogations for particular member states under EU directives which allow member states to proceed towards stated objectives at very different speeds.

How, then, have the different elements of EU energy policy developed?

6.2   Integrating the Internal Energy Market

The Commission's 1988 Internal Energy Market proposals sought to remove barriers to competition such as state assistance, national preferences in procurement and monopoly (Commission of the European Communities, 1988). However, the emphasis of the policy has seemed rather skewed. On the supply side, they did not directly address nuclear power, even though that industry owed its existence to state assistance throughout the fuel cycle. On oil, they sought to equalise specifications for oil products and levels of Value Added Tax and of excise duties across the EU, even though member states were reluctant to cede fiscal sovereignty (certainly not in advance of monetary union), and measures to open up oil exploration and production (see below). Despite the dominance of oil multinationals and State oil corporations, the implicit assumption was that the oil sector was competitive (following a series of initiatives against oil distribution monopolies in the 1970s and 1980s). Moreover the emphasis on energy supply markets meant that they did not address "market failures" on the demand side, despite the many hindrances to the efficient use of energy.

The 1988 proposals benefited from the momentum at the time to complete the internal market and drew upon principles established in the Treaty of Rome, including those on free movement of goods and services and on competition. Since their purpose was to reinforce market forces, they inevitably emphasised short-term considerations and gave no scope for long-term objectives such as supply security, energy conservation and the environment. All of these measures would have required market intervention which would interfere with competition. Nevertheless, there have been a number of important decisions and developments. The major component of the original Single Market programme which affected the energy sector was the procurement liberalisation programme (nearly 20 years after initial efforts to address this problem). This policy aimed to end nationalistic procurement for capital projects in the utility industries by requiring competitive procedures to be open to suppliers from all member States, and the use wherever possible of European specifications (Council of the European Communities 1990b). Competitive tendering now applies to electricity and gas capital projects, and it has been extended to the granting of oil and gas exploration and production licences (Commission of the European Communities 1992b). The most dramatic effect of competitive procurement has arguably been the rapid restructuring of the European power plant industry into a few trans-national groups competing across the EU, bringing to an end the long history of 'national champion' suppliers with protected home markets and surplus heavy electrical manufacturing capacity within the EU as a whole (Thomas and McGowan, 1990).

The first measures to be initiated in the wake of the 1988 paper on the Single Energy Market concerned the pricing and trading practices of the utilities. The 1991 Price Transparency Directive gave industrial users of gas and electricity access to aggregate data on the prices offered by utilities across the EU, a move designed to help industrial users negotiate more effectively with the gas and electric utilities (Council of the European Communities, 1990a). Directives on the Transit of Gas and Electricity required the utilities to carry gas or electricity across their own transmission systems on behalf of other buyers and sellers in adjacent Member States

(Council of the European Communities, 1990c and 1991). The main aim of these Directives was to allow utilities in member states on the periphery of the EU to obtain access on fair and reasonable terms to natural gas from Norway, Russia and other sources and to relatively low-cost electricity generated anywhere in Europe. The major electric and gas utilities say that they had never opposed any transit proposals, but the fact remains that transit has been very slow to take place in the EU and that it has been discouraged by the absence of a requirement upon utilities in member countries positively to support transit proposals.

An accompanying measure to those on transit and transparency was a revision of older European legislation to coordinate investments in the energy sector. The proposal was however abandoned by the Commission. Nonetheless, prior to and since the Internal Energy Market proposals, the EU has been active in the area of supporting investment. For many years, the Commission has earmarked funds from the European Regional Development Fund and the European Investment Bank to finance infrastructure developments, including major electricity and gas interconnectors between the outlying Member States (such as Ireland, Portugal and Greece) and established networks of the 'core' countries (Maniatopoulis, 1992). More recently it has sought to facilitate the development of "Trans-European Networks" (in energy as well as transport and telecommunications) for the economic and social integration of the EU. The EU funding, while modest by comparison with expenditure devoted to other sectors (e.g. transport) takes account not only of the internal profitability of the projects but also their wider economic and social benefits including increased diversity of supply sources and competition, improved load management, and reduced reserve capacity on power systems. Like the Transit Directive, the Trans-European Network initiative is primarily long-term and strategic in character.

Beyond these measures, however, attempts by the Commission to remove barriers to competition have proved harder to pursue. Abolishing coal subsidies and promoting competition in gas and electricity through Third Party Access (TPA) to established transmission systems were crucial parts of the Commission's 1988 Single Energy Market proposals. Both have proved to be very controversial and progress has been limited.

The proposal to ban state assistance for indigenous coal contrasts with the 1951 Treaty setting up the European Coal and Steel Community (the founding Treaty of the EU), which allows coal subsidies in certain defined circumstances. Of more immediate importance to those directly affected, the proposal threatened the very survival of the high-cost German and Spanish coal industries. Despite strong pressure from the Commission, little change occurred because neither country was willing to abandon industries which still remain politically sensitive. By contrast, the Commission seemed powerless to stop or slow down the closure of many relatively low-cost pits in Britain, where the Government did not generally speaking subsidise the coal industry and where the market for indigenous coal had shrunk rapidly due mainly to the effects of electricity privatisation. For a time the Commission envisaged a reference price system to help concentrate EU coal production on lower-cost capacity, but this had no support from Germany and Spain which did not want to close their pits, nor from Britain which did not want to subsidise theirs. Faced with stalemate, the Commission agreed to allow Member States to protect up to 20% of the fuels used in electricity generation. This allowed the continuation of the coal subsidies in Germany and Spain as well as permitting the British Government to subsidise its high-cost nuclear stations.

For national political reasons which were no fault of the Commission, coal production in the EU has thus been rationalised not on the lower cost pits but on the higher cost pits. Due to this, coal prices in Germany and Spain are particularly high, while electricity prices in England and Wales from 1990 to 1995 were 8-10% higher than necessary due to the nuclear subsidy. (The fossil fuel levy was announced as necessary to meet the huge financial liabilities resulting from past nuclear investment. Since much of the money received was actually used to finance construction of Sizewell B, it is reasonable to regard the levy as a subsidy).

The other seemingly intractable problem concerns the Commission's proposal to promote competition by requiring Third Party Access (TPA) to gas and electricity transmission systems. This was intended to give choice to customers normally denied it by the natural monopoly inherent in established integrated transmission networks and the territorial monopolies granted to gas and electricity distribution companies. The aim was to ensure access to the use of transmission systems on non-discriminatory terms by suppliers with gas or electricity to offer but who did not own the transmission system. These suppliers might include entirely new marketing companies with no previous gas or electricity experience as well as major established utilities keen to export their surpluses to or across neighbouring countries.

Although the Commission raised the issue of TPA in 1988, it became clear that the principle was opposed by most of the industry and most member states and it was not until 1992 that the Commission published a draft proposal on the issue. This initially limited access to around 500 large industrial users and large distributors of electricity and gas, although this was to be followed four years later with a comprehensive TPA scheme to apply to all distributors and all final consumers (Commission of the European Communities, 1992a). This proposal was amended in 1993 to allow for a negotiated scheme of TPA as opposed to a mandatory and comprehensive scheme (Commission of the European Communities, 1993). The new proposal led in turn to a proposal by the French for a 'single buyer' scheme whereby Electricité de France or major electric utilities in other countries would be the sole agent for arranging and transmitting electricity imports into their home markets. Since this would have left existing monopoly powers intact, this proposal was widely seen as fundamentally incompatible with EU competition policy (although a final Commission ruling is still pending at the time of writing). TPA appears to be no nearer and, like various other policies, it may have to await the future constitutional development of the EU.

So what went wrong? There were both general reasons which are likely to apply whenever comprehensive TPA is proposed in the future, and specific factors which may not apply with equal force in future.

The pattern of utility ownership, organisation and functions in Europe is varied and complex. Whether publicly-owned and centralised, and whether vertically integrated or specialised transmission or distribution companies, electric utilities in Europe tend to have considerable political support from their national, regional or municipal governments.

Apart from recent experience in Britain due to electricity and gas privatisation which resulted in industry structures quite unlike those in other EU member countries, there is no experience of TPA in the EU. Although there was strong resistance from the vested interests involved, there were also genuine concerns that widespread competition via electricity and gas networks would lead to 'stranded assets' (inability to recover large capital outlays on existing power stations or sell all the gas under long-term 'take-or-pay' contracts), inability to finance new power station construction or take on new long-term gas contracts, and hence reduced security of supply. There were also concerns that it might erode the traditional universal service obligation (the requirement to maintain supplies to all consumers in a given area, to charge published tariffs, and to avoid 'undue preference' or discrimination for particular types of consumer) (Parker and Surrey 1994).

If TPA was to work satisfactorily, it was clear at the outset that it would need a common framework of regulation and effective scrutiny of transmission and distribution charges (to limit monopoly pricing and cross-subsidies) throughout the EU. But the Commission had no powers to impose a common framework of regulation and it could hardly expect the majority of member states which have been opposed to TPA to set up regulatory agencies of their own to enforce TPA.

The specific adverse factors have included: the post-Maastricht emphasis, especially by the British, on 'subsidiarity' and nationalism over anything which smacked of 'Brussels federalism' - even though the British Government continued to extol the virtues of its own model of electricity and gas competition; the lack until 1998 of a large gas pipeline linking the British and the continental gas markets; and Electricité de France's opposition to TPA on its own system while nevertheless seizing every opportunity to export its surplus nuclear electricity to neighbouring countries. It would seem that progress on electricity and gas competition will probably have to await more propitious political circumstances.

6.3   Environmental Protection

Due to the rise of the Green Party and fears of the effects of acid rain on its forests, Germany imposed strict regulations on SO2 and NOx emissions in the early 1980s. As a result, German electric utilities spent DM 20bn installing flue gas desulphurisation (FGD) equipment at 37 GW of coal-fired power stations. Under pressure from Germany and several other countries, in 1984 the Commission proposed uniform reductions in SO2 and NOx emissions for all Member States. After long negotiations in which the British held out for scientific proof of the ecological effects of acid rain and for the development of clean coal combustion technologies which promised to be more energy efficient and less costly than installing FGD, the Large Combustion Plant Directive was agreed in late-1988 (Council of the European Communities 1988).

This Directive set SO_2 emission reduction targets (on a 1980 base) of 40% by 1993, 60% by 1998 and 70% by 2003 for Germany, France, Belgium and the Netherlands. These countries were comparatively well placed to meet such targets. Germany had already invested in a large FGD programme; in France and Belgium, nuclear power formed around 70% of power generation; and the Netherlands had a high proportion of gas-fired power generation.

Lower targets were set for Britain, Italy, Spain and Denmark in view of their greater difficulty and costs in reducing emissions. But its high proportion of coal-fired generation meant that Britain faced the highest compliance cost because it would need to retrofit 12 GW of capacity with FGD. Once electricity privatisation had intervened, the newly privatised generators cut the FGD programme by 50% and took advantage of cheap gas to install a large programme of gas-fired plant instead. The financial savings were considerable and were made possible by the timely removal of long-standing bans on building gas-fired power stations by the British Government and the European Commission, combined with the competitive development of large volumes of very low-cost gas in the British sector of the North Sea (sold at a beach price of 10p per therm in mid-1995). Also following German measures, the EU has subsequently introduced regulations to reduce emissions from smaller industrial plant and from road vehicles.

The fear of global warming led Germany, the Netherlands and Denmark to adopt policies to reduce CO_2 emissions and to press for similar policies for other EU Members States. The EU accounts for 14% of world energy-related CO_2 emissions compared with 23 % for the USA and 5% for Japan (International Energy Agency 1994). It was recognised that, unless the industrial countries set an example, the developing countries were unlikely to do much to contain their growing CO_2 emissions. The Council of Ministers decided in October 1990 that the EU should aim to stabilise CO_2 emissions by 2000 at their level in 1990 and asked the Commission to draw up proposals. These were presented to the Council in October 1991 and revised in June 1992 (Commission of the European Communities 1992c).

The Commission estimated that, without any countermeasures, EU emissions of CO_2 would increase by 11% during the 1990s and it concluded that the factors which led to the marked improvement of energy use efficiency from 1975 to 1985 (due to high energy prices, the decline of energy-intensive industries, and considerable energy-saving technical change) would be absent or much weaker in the 1990s. There was little scope to switch to nuclear power since there was a de facto nuclear moratorium in most member states, while France and Belgium had so much nuclear capacity that they could not economically install more. Given wide variations between member states in the level of CO_2 emissions per capita and in the breakdown of those emissions by sector and by fuel, the Commission thought it impractical to rely on physical regulations of different emission sources and target reductions for individual member states. In view of the scientific uncertainties surrounding global climate change, the Commission also saw the need for a 'no regrets' strategy, by which it meant policy measures that would produce useful benefits (e.g.. reduced traffic congestion and improved supply security due to higher energy use efficiency), should further scientific research conclude that it is unnecessary to reduce CO_2 emissions.

The approach adopted was to try to improve the efficiency of energy use and to change the fuel mix away from high carbon fuels (coal and oil) towards greater use of gas and renewables. The Commission recommended programmes both to promote the development of energy efficient technologies and renewables and the introduction of fiscal measures to encourage energy use efficiency and a switch away from high carbon fuels. Programmes were put in place to reduce EU emissions of CO_2 by 3-4.5% (SAVE) and to increase renewables from 4% to 8% in the EU primary energy balance by 2000 (ALTENER).

Much more radical and controversial was the Commission's proposed energy/carbon tax. The Commission proposed to introduce the new tax at a level equivalent to $1 per barrel of oil, rising $1 a year to $10 per barrel of oil, with 50% of the tax being levied on the heat content of the fuel and 50% on its carbon content. Renewables would be exempt from both parts of the tax and nuclear energy would be exempt from the carbon part of the tax. Initially it was to have been a carbon tax only, but the addition of the energy component was held to be justified by the need to avoid unfairness among countries with widely different proportions of fossil and nuclear power generation. The staged increases would convey a clear price message to consumers, the energy part of the tax would promote energy use efficiency, and the carbon part of the tax would promote the use of renewables and gas and discourage the use of coal and oil. The tax would apply at the same rate throughout the EU, Member states would keep the considerable revenue expected to arise and they would be expected to reduce other taxes in order to offset the macroeconomic effect of the new tax.

The main purpose of the proposed tax was to influence technical choices and its effect was expected to be long-term. The Commission did not expect the tax to reduce CO_2 emissions by more than 3.5% in 2000, with somewhat larger reductions in industrial and household emissions and lower reductions in power station and transport emissions (Faross and Decker 1992). The latter estimate was unfortunate because transport CO_2 emissions are large and growing relatively fast: even after taking account of the proposed tax, transport-based CO_2 emissions were expected to be 21% higher in 2000 than they were in 1990.

The proposals have come in for much criticism. Some have argued that the measure would be pointless and costly for the EU unless other major industrial countries introduced equivalent measures, imposing a particularly heavy impact on energy-intensive industries (though the Commission agreed to allow temporary exemptions for such industries). Critics also argued that the tax would be ineffective in reducing energy demands which are price inelastic, particularly if the new tax merely substituted for existing national taxes and excise duties on transport fuels, and regressive if the fiscal burden fell disproportionately on the poor (especially if the better-off benefited from reduced direct taxes due to the revenues from the energy/carbon tax, as seemed likely).

Although the EU is committed to reducing CO_2 emissions and to setting an example for developing countries which are likely to cause most of the future growth in CO_2 emissions, it has held back on the energy/carbon tax due to the criticisms, the fear of losing international competitiveness and dissent within the Council of Ministers. The EU's environmental objectives have not been helped in recent years by high unemployment, fears of further loss of Europe's competitiveness and the continuing lack of scientific consensus regarding the inevitability, timing and effects of global climate change.

After several years when it was the centrepiece of EU policy on global climate change, the proposed carbon/energy tax was quietly dropped at an EU Council of Ministers meeting in December 1994 (when it was reportedly blocked by the British). This had two effects. First, it undermined the EU's leadership role in the international process instigated in Rio at the 1992 Convention on Climate Change - especially when a Commission assessment concluded that the EU was unlikely to meet its Rio commitment to stabilised CO_2 emissions at the 1990 level in 2000, and that thereafter CO_2 emissions were likely to rise again (Commission of the European Communities,1994). Secondly, there was nothing to put in the place of the energy/carbon tax. At the EU level both the SAVE and ALTENER programmes had been funded insufficiently to have more than marginal impacts. Although the policy onus had now shifted to the national level, the majority of member countries did not have measures in place to reduce their own CO_2 emissions in 2000 to the 1990 level (the exceptions were the comparatively small countries of Denmark and the Netherlands).

An interim conclusion therefore is that while the environmental aspect of EU energy policy has had some successes, notably in reducing 'acid rain' emissions, so far it has been ineffective in tackling the major environmental challenge of the moment, 'greenhouse gas' emissions. This assessment can, however, only be temporary given the scientific uncertainties surrounding both global climate change and acid rain, the financial difficulties faced in supporting new policies and the more general problems which any environmental protection measures encounter in a period of economic recession and political "Euro-scepticism".

6.4   Supply Security

Some figures help to put the problem of supply security into perspective. The EU currently uses over 1100m tonnes of coal equivalent per year. It imports 80% of its oil, 40% of its gas and 33% of its coal. If EU energy demand were to grow at 2% a year, by 2010 it would be 50% above the 1990 level. Since coal is likely to be constrained by environmental factors and nuclear power by public opposition and high capital and reprocessing costs, EU consumption of gas and oil is likely to grow appreciably (Commission of the European Communities, 1990). Since indigenous oil and gas production is unlikely to expand significantly if at all, the demand growth would mean major growth in imported hydrocarbons with gas imports probably doubling in the next 15 years, subject to availability. This view might be modified somewhat if significant improvement occurred in the EU's energy efficiency; but even maximum effort to expand renewables (which is geared mainly to electricity generation) could probably have only marginal effect in reducing the future level of dependence on imported oil and gas. Therefore the EU's dependence on the major world producers of oil and gas (the OPEC producers and the Former Soviet Union (FSU) countries) will probably increase substantially.

The EU is trying to deal with this prospect in several ways. It maintains relations with the Gulf oil producers through the Gulf Co-operation Agreement and it has had regular bilateral talks with Algeria and other large suppliers. It provides financial assistance to the FSU countries to repair and modernise their energy infrastructure. Since 1990, the Commission has proposed that the EU should operate other aspects of supply security including strategic oil and gas stocks and emergency sharing arrangements, and it has also promoted a European Energy Charter which resulted in 1994 in a Treaty Agreement (Commission of the European Communities 1991, Dore and de Bauw, 1995).

The Charter stemmed from a Dutch proposal in 1990 with the purpose of setting up a framework for East-West co-operation in the energy sector and defining the relationship with the transit countries as well as the producing countries (particularly Russia, given that western Siberia has 40% of world gas reserves). It has involved the negotiation of a Treaty Agreement to establish the necessary legal framework covering mining rights, corporate taxation, profit repatriation and a basis for the settlement of commercial disputes; and it is hoped that it will lead to agreements on specific areas of technology transfer of interest to the East European countries (nuclear power, hydrocarbons, energy efficiency and clean coal technologies). The Charter is a strategic initiative based on the West's interest in ensuring secure and expanding supplies of gas and oil from the East, and the East's interest in ensuring large flows of investment capital, hard currency earnings and key technologies from the West.

The Charter Treaty was signed not only by the EU and the FSU countries but also by numerous other OECD countries, including Japan. But the USA and Canada, despite participating in the negotiations, have not signed the Treaty. The purpose of the Charter still lies in the future, for it has achieved few early tangible results. This is due to several factors: the continuing low international oil and gas prices have given little incentive for large-scale Western investment in Russian oil and gas; the fall in oil and gas consumption in the FSU countries has postponed the need for new production and transmission capacity; the bad economic and political situation and the unreformed fiscal and legal systems of the FSU countries still represent considerable uncertainties and risks for Western companies.

The problem of oil and gas supply security depends on how one judges the political risks. The current conventional wisdom, like that in the 1960s, seems to be that oil and gas will remain cheap over any credible planning horizon. Ironically, the costs of increasing security of supply are always seen to be astronomic during periods of cheap energy but perfectly acceptable after oil shocks like those of the 1970s. However, nothing has altered the basic truth which was evident during the oil crises of 1953, 1956, 1967, 1973/4 and 1978/9: OPEC controls the great bulk of world relatively low-cost oil supplies and the great bulk of gas resources which might be available to Europe are in the FSU, Algeria, the Middle East and Nigeria. Nothing approaching a common EU foreign policy, nor common military forces, is in prospect. Nor is it likely that the EU will acquire the geopolitical weight which the USA has had over the past 50 years. On oil supply security, the EU member states still look to the International Energy Agency (IEA) because this brings in the USA and other OECD countries which together have much greater weight in the international oil market than the EU has.

Nevertheless, the EU has a large internal market and the major oil and gas exporters have a strong interest in maintaining the benefits of trade, foreign investment and technical collaboration. Compared with reliance on military power, Europe's interests are likely to be better served by building upon mutual commercial interests with the producers and its historically close ties with the OPEC countries. As far as supply security matters are concerned, there is a question as to whether the EU should speak with one voice in IEA deliberations.

6.5   Conclusion: The Scope for EU Energy Policy

A realistic assessment of the scope for EU energy policy must start with the facts of the situation. These appear to be as follows:

- There are obvious conflicts between the three basic policy objectives. The conflicts are acute during periods, like the present, of low energy prices.

- The European Treaties are mainly concerned with competition and trade: they do not give the EU special powers over energy matters and they are inconsistent with the need to intervene to prevent market forces from having ill effects.

- Views are divided on the future constitutional development of the EU between a process of 'deepening' via monetary union to full political union or of 'widening' through the admission of new member states to create a much looser association which would be little more than a large free trade zone.

- The 'minimalist' view of the role of the state in economic affairs taken by recent Conservative British Governments conflicts with both the German-style social market economy (e.g.. as reflected in the German coal subsidy) and with French-style dirigism characterised by nationalism, technocracy and elitism. These differences could obviously affect national concepts of the content and instruments of the energy policies conducted by member states as well as those of common policy at the EU level.

- Member states have always pursued separate policies on energy supply security, reflecting differences in national viewpoint, in indigenous fuel resources, and in the protection and promotion of national fuel and plant suppliers. This is in strong contrast with environmental protection and with research and development for which the EU has acquired considerable authority.

- Member states tend to see the International Energy Agency (IEA) rather than the EU as the appropriate body to handle international oil emergency arrangements.

- In most EU member countries, nuclear power construction is virtually ruled out by the reduced acceptability of nuclear power to the public since Chernobyl and the rising costs of nuclear generation, while France has so much nuclear capacity that for the foreseeable future it is unlikely to be economic to install any more.

- To the extent that TPA and competition in final markets for electricity and gas increase the risk of 'stranded assets' (or under-recovery of capital costs), they will tend to discourage investment in major new power stations and gas supplies.

On the basis of these factors it might seem logical for the European Commission to do away with its Energy Directorate and leave energy policy to the member states. The powers of the Competition Directorate would be sufficient to ensure that the market rules were respected, whereas, if the Energy Directorate is maintained, it will continue to be faced with the contradiction of having to show that it is acting with regard to energy policy in ways which honour the EU's commitment to market forces which logically must deprive member states of their principal means of action through subsidy, taxation and other forms of market intervention. But abolishing the Energy Directorate would solve little: the problem is that energy policy in the EU is subordinated to wider objectives in EU policy making and the primacy of the 'single market' policy is constraining energy policy decisions unduly, not only at the EU level but also at the national level. And in any case, there are further considerations to be taken into account:

- Conflicting policy objectives are not unusual at the national level or the EU level. What matters is the balance of priorities when decisions are taken. Barriers to internal trade in energy and in energy plant and equipment have been and are continuing to be reduced.

- If and when member states move to political union, more coordination on energy matters will be necessary at the EU level. Meanwhile, there is a sense of common interest in the EU which, in the event of a serious supply or environmental problem, might well lead to more energy cooperation between member states.

- The Commission plays an increasingly useful role in energy efficiency, environmental protection, renewables, technology transfer and energy R&D.

- The objectives of the dialogue with the Gulf oil producers, the European Energy Charter, and the attempt to reduce 'greenhouse' gas emissions are all long-term. It would be sheer 'short-termism' to discount their potential value at this stage.

The EU is gradually taking shape and the energy sector has played its part in the process of integrating national markets into the large internal EU market. Given the historical importance of EU competition policy, how far should the Commission push its market - oriented reforms in the energy sector and should the EU have a Common Energy Policy?

We hope the EU reforms will go to the point at which any remaining government interventions and subsidy must be carefully justified in terms of costs and benefits but not so far that the market-oriented reforms reduce the ability of the fuel industries to plan ahead, coordinate and invest in new capacity. We see little scope for a Common Energy Policy consisting of identical priorities and measures for all member states. The Common Policy could not be well-tailored to each country's needs and it would be too inflexible. We think that the Energy Directorate should be retained and that one of its more important tasks should be to hold discussions with member states to try to define an appropriate division of decision making responsibilities between the EU and the member states, with the latter retaining considerable specific decision making responsibilities. For the reasons given earlier, the EU should have responsibility in areas such as overseas energy relations (including the FSU, the Gulf producers, and Algeria), the economic regulation of electric and gas utilities, control of environmental emissions, the extension of electricity and gas grids to outlying parts of the internal market, internal coordination of fuel stocks and other oil and gas emergency arrangements and presenting a common EU position in other international forums such as the IEA and talks on global climate change. These activities may not constitute a comprehensive Common Energy Policy, but they do offer the EU a constructive contribution to energy policy-making.

References

Commission of the European Communities, The Internal Energy Market, (COM (88) 238), CEC, Brussels, 1988.

Commission of the European Communities, Security of Supply, the Internal Energy Market and Energy Policy, SEC (90) 1248, CEC, Brussels, September 1990.

Commission of the European Communities, A European Energy Charter, COM (91) 36, CEC, Brussells 1991.

Commission of the European Communities, Proposal for a Council Directive Concerning Common Rules for the internal market in electricity and natural gas, COM (91) 548, CEC, Brussels, 1992a.

Commission of the European Communities, Proposal for a Council Directive on the Conditions for granting and using authorisations for the prospection, exploration and extraction of hydrocarbons, COM (92) 110, CEC, Brussels, 1992b.

Commission of the European Communities, Proposal for a Council Directive Introducing a Tax on Carbon Dioxide Emissions and Energy, COM (92) 226, CEC, Brussels, 1992c.

Commission of the European Communities, Amended Proposal for a European Parliament and Council Directive Concerning Common Rules for the Internal Market in Electricity and Natural Gas COM (93) 643, CEC, Brussels, 1993.

Commission of the European Commission, Proposals for Council Decisions concerning the specific programmes implementing the EU framework programmes, COM(94) 70, CEC, Brussels, 1994.

Council the European Communities, Directive of 24th November 1988 on the Limitation of Emissions of Certain Pollutants into the Air from Large Combustion Plants, L336/88, OOPEC, Luxembourg, 1988

Council of the European Communities, Directive of 29th June 1990 concerning a Community procedure to improve the transparency of gas and electricity prices charged to end users, L 185/90, OOPEC, Luxembourg, 1990a.

Council of the European Communities, Directive of 17th September 1990 on the Procurement Procedures of Entities operating in the Water Energy Telecommunications and Transport Sectors, L297/90, OOPEC, Luxembourg, 1990b.

Council of the European Communities, Directive of 29th October 1990 on the Transit of Electricity through transmission grids, L313/90, OOPEC, 1990c

Council of the European Communities, Directive of the 12th June 1991 on the Transit of Natural Gas through Transmission Grids, L 147/91, OOPEC, Luxembourg, 1991.

Dore, J and de Bauw, R The European Energy Charter, RIIA, 1995

Faross, P and Decker, M "CO_2 Emissions Stabilisation: the Community Strategy" Energy in Europe, CEC, Brussels, 1992.

International Energy Agency, Climate Change Policy Initiatives: 1994 Update.

Maniatopoulis, C "European Community Energy and Natural Gas Policies", Energy in Europe, CEC, Brussells, 1992.

Parker, M and Surrey, J "UK Gas Policy: Regulated Monopoly or Managed Competition?" STEEP Special Report, University of Sussex, 1994.

Thomas, S and McGowan, F The Market for Heavy Electrical Equipment, NEI, 1990.

Printing: Weihert-Druck GmbH, Darmstadt
Binding: Theo Gansert Buchbinderei GmbH, Weinheim